Resilient Communities of Central

This book argues for the need to rethink governance through the lens of 'resilience as self-governance'. Building on complexity-thinking, it contends that in the context of change and complex life, challenges are most efficiently dealt with, at the source, 'locally', to make 'the global' more responsive and sustainable.

Resilience as self-governance is advanced as an overriding framework to explore its constitutive elements – identity, 'good life', local coping strategies and support infrastructures – which, when mobilized, can turn communities into 'peoplehood' in the face of adversity. It is argued that these *communities of relations*, self-organised and self-aware of their worth, is what makes them so *resilient* to crises, and what helps them to transform with change; and how they should be governed today. Central Eurasia, spanning from Belarus in the west, to Azerbaijan in the south and Kyrgyzstan in the east, provides fertile grounds for exploring how resilience works in practice in times of complex change. By immersing into centuries-long traditions and philosophy, local experiences of survival, and visions for change, this book shows that governability at any level requires a substantive 'local' input to make 'the global' more enduring and resilient in a complex adaptive world.

This book will be of great value to students and scholars in the fields of Politics including Eurasian politics and the various aspects of Governance. Most of the chapters in this book were published as a special issue of *Cambridge Review of International Affairs*.

Elena Korosteleva is Professor of Politics and Global Sustainable Development, and Director of the Institute for Global Sustainable Development, at the University of Warwick. Elena is formerly Principal Investigator for GCRF-funded project COMPASS (ES/P010849/1, 2017–22) and Co-Founder/Investigator for the Oxford Belarus Observatory (2020–22), University of Oxford. Her interests include resilience, complexity-thinking, order formation and multi-order governance in Central Eurasia. Her recent publications are *Belarus in XXI Century: Between Dictatorship and Democracy* (with I. Petrova and A. Kudlenko, Routledge 2022, forthcoming), 'The War in Ukraine: Putin and the Multi-order World', *Contemporary Security Policy*, 43(3) 2022: 466–81 (with T. Flockhart); and *Resilience in EU and International Institutions* (with T. Flockhart, Routledge 2020).

Irina Petrova is Assistant Professor in the Politics of Eurasia at the UCL School of Slavonic and East European Studies (SSEES). Her recent publications include 'Community Resilience in Belarus and the EU response' in *Journal of Common Market Studies Annual Review,* October 2021 (with E. Korosteleva); 'Societal fragilities and resilience: The emergence of peoplehood in Belarus' (with E. Korosteleva), August 2021 in the *Journal of Eurasian Studies;* and 'From "the global" to "the local": the future of cooperative orders in Central Eurasia in times of complexity' (with E. Korosteleva), *International Politics* 58(3) 2021.

Resilient Communities of Central Eurasia

Responding to Change, Complexity and the Visions of 'The Good Life'

Edited by
Elena Korosteleva and Irina Petrova

LONDON AND NEW YORK

First published 2023
by Routledge
4 Park Square, Milton Park, Abingdon, Oxon OX14 4RN

and by Routledge
605 Third Avenue, New York, NY 10158

Routledge is an imprint of the Taylor & Francis Group, an informa business

Chapters 2, 6 and 7 © 2023 Department of Politics and International Studies
Chapter 5 © 2023 Global South Ltd
Chapter 3 © 2023 Emilian Kavalski
Chapter 4 © 2023 Nargis Nurulla-Khojaeva
Chapter 8 © 2023 Chiara Pierobon and Zarina Adambussinova
Introduction © 2022 Elena A. Korosteleva and Irina Petrova. Originally published as Open Access.
Chapter 1 © 2021 David Chandler. Originally published as Open Access.
Conclusion © 2021 Prajakti Kalra. Originally published as Open Access.

British Library Cataloguing in Publication Data
A catalogue record for this book is available from the British Library

ISBN13: 978-1-032-29094-2 (hbk)
ISBN13: 978-1-032-29095-9 (pbk)
ISBN13: 978-1-003-29999-8 (ebk)

DOI: 10.4324/9781003299998

Typeset in Baskerville BT
by Newgen Publishing UK

Publisher's Note
The publisher accepts responsibility for any inconsistencies that may have arisen during the conversion of this book from journal articles to book chapters, namely the inclusion of journal terminology.

Disclaimer
Every effort has been made to contact copyright holders for their permission to reprint material in this book. The publishers would be grateful to hear from any copyright holder who is not here acknowledged and will undertake to rectify any errors or omissions in future editions of this book.

Contents

Citation Information

The following chapters, except for chapter 5, were originally published in the journal *Cambridge Review of International Affairs*, volume 35, issue 2 (2022). Chapter 5 was originally published in the journal *Central Asian Survey*. When citing this material, please use the original page numbering for each article, as follows:

Introduction
What makes communities resilient in times of complexity and change?
Elena A. Korosteleva and Irina Petrova
Cambridge Review of International Affairs, volume 35, issue 2 (2022), pp. 137–157

Chapter 1
Decolonising resilience: reading Glissant's Poetics of Relation *in Central Eurasia*
David Chandler
Cambridge Review of International Affairs, volume 35, issue 2 (2022), pp. 158–175

Chapter 2
From 'Westlessness' to renewal of the liberal international order: whose vision for the 'good life' will matter?
Trine Flockhart
Cambridge Review of International Affairs, volume 35, issue 2 (2022), pp. 176–193

Chapter 5
Communal self-governance as an alternative to neoliberal governance: proposing a post-development approach to EU resilience-building in Central Asia
Fabienne Bossuyt and Nazima Davletova
Central Asian Survey, DOI: 10.1080/02634937.2022.2058913

Chapter 6
Belarus between West and East: experience of social integration via inclusive resilience
Victor Pravdivets, Anna Markovich and Artsiom Nazaranka
Cambridge Review of International Affairs, volume 35, issue 2 (2022), pp. 194–209

Chapter 7

The Azerbaijani resilient society: explaining the multifaceted aspects of people's social solidarity
Azer Babayev and Kavus Abushov
Cambridge Review of International Affairs, volume 35, issue 2 (2022), pp. 210–234

Conclusion

Locating Central Eurasia's inherent resilience
Prajakti Kalra
Cambridge Review of International Affairs, volume 35, issue 2 (2022), pp. 235–255

For any permission-related enquiries please visit:
www.tandfonline.com/page/help/permissions

Notes on Contributors

Kavus Abushov is Associate Professor of Political Science at ADA University. His research focuses on ethnic civil wars and intra-state conflicts, security studies and state-building. His geographic area of focus is the post-Soviet space. He earned his PhD in Political Science from the University of Muenster in Germany and did his postdoc at the Massachusetts Institute of Technology.

Zarina Adambussinova is a Social Anthropologist and a Postdoctoral fellow of the Volkswagen Foundation in the Department of Anthropology, Technology and International Development at the American University of Central Asia in Bishkek, Kyrgyzstan. Her research interests include heritage, heritage and memory practices, post-Soviet Central Asia, mono-industrial towns, and informal economic practices. Since October 2020, she has been working on her postdoctoral project called "Dealing with Uncertainty: Socio-Economic Survival Strategies of Local Residents in post-Soviet mono-industrial Towns in Central Asia (Kazakhstan and Kyrgyzstan)".

Azer Babayev is Assistant Professor of Political Science at ADA University. His research interests are in the areas of conflict resolution and democracy promotion, with a focus on the post-Soviet space. Prior to joining ADA University, he was a postdoctoral research fellow at the Peace Research Institute Frankfurt.

Fabienne Bossuyt is Associate Professor at and co-coordinator of the Ghent Institute for International and European Studies (GIES). Her research interests include External policies of the EU; EU relations with Central Asia; and EU relations with the post-Soviet space. She is co-director of the Eureast Platform of Ghent University. She is also a Professorial Fellow at UNU-CRIS and is active as an affiliated researcher of EUCAM and a member of the Academic Board of the European Neighbourhood Council.

David Chandler is Professor of International Relations, University of Westminster, UK. He edits the journal *Anthropocenes: Human, Inhuman, Posthuman*. His recent monographs include *Anthropocene Islands: Entangled Worlds* (with Jonathan Pugh, 2021); *Becoming Indigenous: Governing Imaginaries in the Anthropocene* (with Julian Reid, 2019) and *Ontopolitics in the Anthropocene: An Introduction to Mapping, Sensing and Hacking* (2018).

Nazima Davletova is Adjunct Professor at Webster University, Tashkent, Uzbekistan. She is a senior expert on gender policy (Agency of Information and Mass Communications), and lecturer at the University of World Economy and Diplomacy (UWED). For several years, she worked in think tanks for international relations, and then worked in media relations. She has published several political notes on Russian politics in Central Asia and the media in Uzbekistan.

Trine Flockhart is Professor of International Relations and Director of the Center for War Studies (Policy and external affairs) at the University of Southern Denmark. Her research focuses on international order and transformational change, NATO, European Security, the liberal international order (and its crisis), resilience and transatlantic relations.

Prajakti Kalra is Affiliated Lecturer in the Centre of Development Studies, University of Cambridge and Research Officer in the Cambridge Central Asia Forum, Jesus College, University of Cambridge. She has trained as a historian, political scientist, and a psychologist. Her first book is entitled the *The Silk Road and the Political Economy of the Mongol Empire* (Routledge, 2018). Her research focuses on reimagining the place of Central Asia in world history and its relevance to development processes in the making of modern Eurasia.

Emilian Kavalski is the NAWA Chair Professor at the Complex Systems Lab, Jagiellonian University in Krakow, Poland; and the book series editor for Routledge's 'Rethinking Asia and International Relations' series. His work explores the interconnections between the simultaneous decentering of International Relations by post-Western perspectives and non-anthropocentric approaches.

Elena A. Korosteleva is Professor of Politics and Global Sustainable Development, and Director of the Institute for Global Sustainable Development, at the University of Warwick. Elena is formerly Principal Investigator for GCRF-funded project COMPASS (ES/P010849/1, 2017–22) and Co-Founder/Investigator for the Oxford Belarus Observatory (2020–22), University of Oxford. Her interests include resilience, complexity-thinking, order formation and multi-order governance in Central Eurasia. Her recent publications are *Belarus in XXI Century: Between Dictatorship and Democracy* (with I. Petrova and A. Kudlenko, Routledge 2022, forthcoming), 'The War in Ukraine: Putin and the Multi-order World', *Contemporary Security Policy*, 43(3) 2022: 466–81 (with T. Flockhart); and *Resilience in EU and International Institutions* (with T. Flockhart, Routledge 2020).

Anna Markovich is Senior Lecturer at the Chair of International Journalism of the Faculty of Journalism, Belarusian State University. Her research interests encompass discourse of integration, media research methods and persuasive communications.

Artsiom Nazaranka is Senior Lecturer and Researcher at the Faculty of History, Belarusian State University. His research focuses on History and Practice of Public Administration in Belarus, Data Protection, Migration and EU studies.

Nargis Nurulla-Khojaeva is Professor at Samarkand State University (Uzbekistan). Her main current interest is culture and history in Central Asia. Nargis is a culturologist, an Orientalist, and a philosopher. She studied History and Philosophy at the Tajik State University, named after V.I. Lenin, and her PhD thesis was devoted to gender issues. This subsequently led to significant involvement in the "post-structural" efforts to develop feminist theories in Tajikistan.

Irina Petrova is Assistant Professor in the Politics of Eurasia at the UCL School of Slavonic and East European Studies (SSEES). Her recent publications include 'Community Resilience in Belarus and the EU response' in *Journal of Common Market Studies Annual Review,* October 2021 (with E. Korosteleva); 'Societal fragilities and resilience: The emergence of peoplehood in Belarus' (with E. Korosteleva), August 2021 in the *Journal of Eurasian Studies;* and 'From "the global" to "the local": the future of cooperative orders in Central Eurasia in times of complexity' (with E. Korosteleva), *International Politics* 58(3) 2021.

Chiara Pierobon is currently DAAD Visiting Professor at the Henry M. Jackson School for International Studies and at the Department of Political Science at the University of Washington, Seattle (USA) and Editor at the OSCE Academy in Bishkek (Kyrgyzstan). She received a binational Ph.D. in Sociology and Social Research from the University of Bielefeld (Germany) and the University of Trento (Italy). Her main areas of expertise are EU's support to civil society in Central Asia, prevention of violent extremism (PVE), sustainable development and local ownership, public diplomacy, social capital and resilience.

Victor Pravdivets is Principal Researcher at the Center for Sociological and Political Research of Belarusian State University. His research interests comprise sociology of values, sociology of mass media, methodology of sociological research and environmental sociology.

Introduction: What makes communities resilient in times of complexity and change?

Elena Korosteleva and Irina Petrova

Abstract *This introduction to the Special Issue/edited volume problematizes the necessity to rethink governance through the lens of resilience and suggests a novel conceptualization of resilience. Building the argument on complexity-thinking, this issue contends that in the context of change and complex life, challenges are most efficiently dealt with, at the source, 'locally', to make 'the global' more sustainable. Accordingly, the concept of resilience as self-governance is advanced in the introduction as an overriding framework to explore its constitutive elements – identity, 'good life', local coping strategies and support infrastructures – which, when mobilized, can turn community into 'peoplehood' in the face of adversity. This conceptualization, we argue, explains what makes communities adapt and transform, and how they should be governed today. Central Eurasia, spanning from Belarus in the west, to Azerbaijan in the south and Kyrgyzstan in the east, provides fertile grounds for exploring how resilience works in practice in times of complex change. By immersing into centuries-long traditions and philosophy, local experiences of survival, and visions for change, this introduction – along with the Special Issue/edited volume – shows that governability at any level requires a substantive 'local' input to make 'the global' more enduring and resilient in a complex adaptive world.*

Introduction

The last decades of the 20th century and the beginning of the 21st century have been widely characterised by a 'post'-prefix – e.g., we live in a *post*-modern context, when liberal international order is evolving into *post*-liberalism, when *post*-colonialism and international development are challenged by *post*-development, and anthropocentrism is called into question by *post*-humanism. The 'post'-prefix typically draws on a cumulative knowledge system to deal with change and gaps in learning; but it is also a sign of ongoing transformation, to *retrospectively* rectify the insufficiencies of this knowledge system, while filling those gaps. Indeed, at the turn of the century a range of trends explicitly manifested themselves, leading to a clear understanding that the world is entering a new historical phase: the rise of the post-industrial economy and society, increasing levels of globalization coupled with regionalization and evolving notions of sovereignty, unprecedented inter-connectedness and transnationalization, emergence of a new world order and global challenges of a planetary scale. Reinforcing each other, these developments arguably signified the arrival of an '*entirely new historical period* ... [in which] many ideas and assumptions dominant for decades are rapidly becoming obsolete' (Mishra 2020; see also Macy 2007).

In contrast to the 'post'-terms, more practicable definitions of the new realities today, tackling head on the inherent insufficiency of knowledge, refer to 'the VUCA-world' (Burrows and Gnad 2017) and a 'complex world' (Kavalski 2007, and in this volume). These terms tease out key features of today's environment such as **v**olatility, **u**ncertainty, **c**omplexity and **a**mbiguity of societal development (VUCA). International life is getting more complex in many respects – a multiplicity of global and local actors, interactions of various networks ('multiplexity'), non-linear developments and emergence processes, often through self-organization, in the context of deep interconnectedness, result in a dynamic entanglement, associated with increasing levels of unpredictability and a lack of control (Bousquet and Geyer 2011; Bousquet and Curtis 2011). In fact, the world as we see it today has become far beyond '*post-knowing*' – that is, radically shifting our understanding of it from 'knowing the knowns' with a solution for everything, through 'knowing the unknowns' with few templates to tackle uncertainty; to finally recognizing that full 'knowing' of a complex world is impossible, including a human effort at long-term forecasting and control (Vogelsang 2002; Dooley 1997).

In this context, traditional modes of top-down governance become less relevant or effective for that matter. Indeed, key international programmes – including international development, democratization, and the fight against global warming, poverty, famine and a ravaging health pandemic – have yielded limited and highly controversial results (Edkins 2019). As a solution to increasing complexity, the discourse of resilience entered the narrative and practice of the major international institutions (UN, World Bank, OECD, EU) about three decades ago. The resulting approach focused on building institutions and structures facilitating resilience, understood as an ability of a system to bounce back after crises (Bourbeau 2018). This implied that a problem can be solved locally, yet through 'outside-in' international cooperation premised on the local appropriation of Western templates and resources. Essentially, since resilience as a governance narrative initially emerged in the language of international institutions, its practice as well as its academic reflection have been largely Western-centric (Rouet and Pascariu 2019; Cusumano and Hofmaier 2020). Resilience, therefore, has been amply conceptualised in the literature as a neoliberal practice of governmentality (Walker and Cooper 2011; Zebrowski 2013; Joseph 2013; 2018) targeted at the identification of potential vulnerabilities to be preventively addressed through 'capacity-building', 'empowerment' and the construction of a 'neoliberal subject' (Chandler and Reid 2016).

A more recent line of thinking, attempting to go beyond neoliberalism, defines resilience as 'a new art of governing complexity' (Chandler 2014; 2020; Korosteleva and Flockhart 2020), which shifts the focus from vulnerabilities, adaptation and intervention to transformation and self-governance of 'the local' and the 'problem at source'. This line of thinking argues that communities have capacities and coping strategies that are more attuned to resolving the problems on the ground, with external support as necessary – thus constituting an 'inside-out' perspective (Korosteleva 2018; Juncos 2017). This means that resilience is more about understanding and facilitating these local self-reliant and self-organizing practices, and indeed closer to the 'right to opacity' (Chandler in this volume), rather than about adopting 'modernizing' top-down techniques through inter-national intervention (Finkenbusch 2021). This is not to argue that 'the local' is ideal, and existing practices need to be conserved as they are. Rather, it is argued that a better understanding of resilience as vested in local communities enables more sustainable orders and responsive governance on all levels. These bottom-up and horizontal engagements would make global governance potentially more responsive to change, and indeed 'fit-for-purpose' (see Flockhart in this volume).

In line with this approach, we have defined and explored *resilience* elsewhere (Korosteleva and Flockhart 2020) as both a *quality* of a complex adaptive system[1] and a new *analytic of governance* for an increasingly complex world. We also argued that 'the global' in a complex and unpredictable world cannot be under-stood and managed without 'the local' and 'the person', because it is precisely the intra- and inter-relations of the latter, in their diversity, that come to define the configurations and prospects for sustainability of the global system (Korosteleva and Petrova 2020). In this introduction, and the Special Issue more broadly, and as a next step of our inquiry into resilience,[2] we aim to unpack it further as 'the local', this time, however, through the lens of its core constitutive elements – e.g., identity shaped and driven by a sense of a 'good life'; infrastructures of communal support; philosophy and traditions of neighbourliness; solidarity and convoca-tion of the peoplehood (Korosteleva and Petrova 2021) – as *a process* that makes communities endure and transform in the face of adversity. Understanding how resilience as self-governance works in practice may give us a better sense of what kind of multi-level governance is needed to make an entangled, complex and perceivably more hostile world – ridden with global challenges and crises – more responsive and adaptive to change.

This edited volume develops synergies between different ways of thinking and practices (including their geographical and epistemological variations), and substantially reshapes our understanding of resilience, community, change, and governance – key concepts of International Relations (IR). Most notably, by unpacking the workings of resilience through the lens of local communities, this Special Issue contributes to a better understanding of change and complexity, and our response to it, in a search for more sustainable models of governance on all levels. We place our discussions into a particular geographical focus, which, following Scott Levi (2020: xiv; Korosteleva and Paikin 2020), we call *wider Eurasia*, or *Central Eurasia*, by which we mean 'the full Eurasian interior' (Levi 2020: viv), embracing Eastern Europe, the Caucasus and Central Asia. This denomination seeks to specifically avoid treating it as a homogenous region, and instead conceive of it as an expansive *locality*, covering a wide area of diverse cultures, traditions and thinking, that is nonetheless unified by a sense of common 'lived' history, and its inter-generational legacies. Building on these dispersed and yet embedded practices and 'memories' of 'living', 'surviving' and 'transforming together', we are hoping to capture strategies of resilience, both historically defined and contemporary manifested, drawing on local aspirations, practices, and philosophies. Most notably, these include a sense of solidarity and good neighbourliness reflected in the enduring notions of '*hamsoya*' (sharing a shadow); '*baghdad al wujad*' (unity of beings); '*hamdardi*' and '*ham-dili*' (compassionateness, kindness and forgiveness); and much more (Nurulla-Khojaeva in this volume). These form an important mesh and poetics of relations (Glissant 1997), order and organisation, which enables people to strive together for a *life worth living*, and to stand tall as a community in the face of adversity. This volume was born out of a series of conferences and workshops, held across Eurasia, assembling scholars of a pluridisciplinary background – IR, politics, sociology, anthropology, history, physics, and culture – who focus on resilience both theoretically and empirically, being both 'international' and 'local', but invariably part of the UK Global Challenges research network COMPASS.[3]

The introduction will proceed by first contouring complexity as a new reality of *post-knowing* and a framework for understanding the VUCA-world, introducing *resilience* as a new analytic of governance for managing complexity, bottom-up and inside-out. It will then unpack 'resilience' as a complex *assemblage* 'where relations [being exterior to their terms] are the understanding of the contingent emergent effects of interaction' (Chandler 2018, 63, with reference to the work of Gilles Deleuze and Felix Guattari) of constitutive elements, as well as *a process of becoming with*, to understand their meaning and relationality, before they are explored empirically in the Special Issue through the manifold locality of Central Eurasia as a rich and heterogeneous space. The introduction will also explain the relevance and poignancy of Eurasia as a focal geography for this discussion; and premise the volume's contributions by threading them together into a complexity-framed argument positing resilient communities as a gateway to a more cooperative and sustainable multi-governance and multi-order world (Flockhart 2016).

Complexity, and resilience in the VUCA world
In addition to the transformational change making the world more pluriversal, unstable and unpredictable than ever before, another important ongoing transformation is of an epistemological nature. The way we think about the world has changed drastically in the course of the 20th century. The principles of uncertainty and unpredictability of the quantum world (Gell-Mann 1995), the theory of relativity, the challenges to the Darwinian worldview which commingles both self-organisation and selection (Kauffman 1995) overturned the deeply-entrenched Newtonian/positivist thinking in natural sciences, conceiving of the world and the universe as based on universal laws waiting to be discovered, and pushed us, in the words of Latour, to face Gaia, 'the grand inhibitor of circular thinking, and a great impetus to thinking outside the box' (2017, 6). Mesh[4] understandings of the universe (Kurki 2020) humbled us and highlighted the limitations of our possibility of knowing and understanding; and yet, it is precisely through the

realization of these limitations that we slowly begin to feel 'at home in the universe' while searching for and internalizing the principles of complexity (Kauffman 1995). Over the past few decades, complexity-thinking has been proliferating in social sciences and became embedded in a number of theories, as will be discussed below. This double change – of the world we live in and the way we understand it, by facing Gaia – requires a profound rethinking of International Relations, as argued in the next chapter by Kavalski.

In this context, complexity-thinking offers a more optimal conceptual lens to analyse society and international affairs. In what follows, we discuss the main assumptions of complexity-thinking and its implications for governance studies and International Relations. We argue that resilience as a quality of a complex system and an analytic of governance (and self-governance) is emerging as a response to complex life and the inability to govern in a habitual top-down way. Drawing on these insights, we advance the argument further by unpacking what makes communities more resilient and re-connecting this local perspective back to 'the global' for more sustainable international orders and more responsive governance on all levels, as argued, for example, by Kalra and Flockhart in this volume.

The logic of complexity-thinking

Complexity-thinking lies at the heart of conceptualizing resilience both as a quality of a complex (adaptive) system and as an analytic of governance. Hence, before unpacking the elements that constitute resilient communities, it is essential to summarize the main tenets of the underlying theoretical framework. Complexity-thinking originates in the challenge posed by the 'uncertainty principle' developed in quantum mechanics of the Cartesian scientific paradigm dominant since Enlightenment (Jørgensen 1990). Heisenberg's 'uncertainty principle' proved that 'at the quantum level of tiny particles it was impossible to measure both mass and momentum simultaneously, making access to full information impossible' (Chandler 2014, 48, and in this volume). This discovery marked a breakaway with the belief that natural and social laws can ultimately be uncovered. Instead, it advanced a new epistemology based on the premises of complexity postulating limitations of scientific knowledge.

Complexity is akin to systems-thinking in that it differentiates between simple/closed and complex/open systems. In contrast to simple systems, where the outcome is causally determined by a set of inputs, complex systems cannot be meaningfully understood based on the analysis of their parts. This is because complex systems consist of a vast constellation of different types of actors connected into (often heterogeneous) networks, which are, in turn, related to one another. Furthermore, complex systems are characterised by non-linearity, i.e., processes in which change in inputs is not proportional to the change of the output. Relational links among the elements of a system therefore become essential, as a tiny change through a chain of interconnections and adaptations may result in substantial output variation, commonly known as a 'butterfly effect'

(Eoyang and Berkas 1998, 7). For that reason, complexity-thinking implies thorough relational and processual analysis, adding value to the existing debates already raised on the pages of this journal[5] and elsewhere. Many processes in complex systems are emergent, aiming for a system equilibrium through a series of iterative adaptations. Emergence is therefore defined as 'the fact that the individual interaction level produces social effects at the macro level, which are not reducible to the aggregate alone' (Schneider 2012, 138; see also Holland 1995). Hence, the central idea is that collective and cooperative orders develop from below and horizontally as a result of self-organization, requiring no central control (Kauffman 1995).[6]

The meaning of resilience

Complexity-thinking is best suited to 'semi-turbulent and turbulent environments where change is imminent and frequent' (Dooley 1997:92), where the realities of the VUCA-world we are facing today urgently demand such an epistemology. Complexity-thinking accounts for *self-reliance*, and collective *self-organization* in the face of adversity, which in turn draw on 'a *shared vision*' of *becoming with* (Berenskoetter 2011; Chandler in this volume) and 'individual's readiness for change' (Berenskoetter 2011, 91), as well as *inherent* communal resources, processes and *capacities*, because all fundamental forces and structures 'arise from *local processes* and not by means of action at a distance' (Gell-Mann 1995, 177).

   All these core tenets of complexity-thinking ensure a most optimal response to emergence and change, and as we argue elsewhere (Korosteleva and Flockhart 2020, and Kavalski in this volume), are quintessentially reflected in the concept and practice of *resilience*. In the official discourse of the European Commission, resilience is defined as 'the inherent strength of an entity – an individual, a household, a community or a larger structure – to better resist stress and shock, and the capacity of this entity to bounce back rapidly from the impact' (2012:5). This SI, however, proposes that resilience is not just a *quality of a complex (adaptive) system* that enables entities to respond more adequately to change in search of an equilibrium (Korosteleva 2020; Luhmann 1990). In the context of complex life, resilience 'is always more' (Bargués-Pedreny 2020) and should also serve as an *analytic of governance* both in terms of thinking and the practice of governing. In this sense, resilience becomes a reliable operational tool for complex governance to function effectively, because it allows for more responsive and sustainable *governability* – bottom-up and relational (Kurki 2020; Manson 2001) – and *futuring*. In a complex world 'there may not be a predictable future but there is still a need to engage in futuring' (Vogelsang 2002, 10) by continually constructing the future bottom-up and horizontally. Moreover, given the 'mesh' or dynamic entanglement ontology that we adopt, following the footsteps of Morton (2010; 2013) and Kurki (2020), adaptation is not enough for resilience. To be genuinely resilient, communities must be able to transform with change, for the system is in constant flux.

In a complex world where 'random local rules of behaviour can result in emergent order at a global level' and where 'whether there is order or not depends upon the degree of connectedness between the elements of the network' (Stacey 1995, 488), resilience emerges as a useful framework to explore the role of 'the local' in shaping 'the global' through connectedness and ensuring its adaptability to change. By 'the local' we here refer to the person, the community or the society, which we loosely term here as 'community of relations'. By analysing the local and its importance for responding to global challenges, this volume contributes to the burgeoning literature emerging at the intersection of political philosophy (Chandler et al. 2020; 2021; Clark and Szerszynski 2021), IR (Kavalski 2007; Bousquet and Geyer 2011; Acharya and Buzan 2017; Qin 2018; Nordin et al. 2019; Reus-Smit 2018; Kurki 2020), governance and design (Escobar 2018; Kothari et al. 2019), EU studies (Keukeleire et al. 2020; Fisher-Onar and Nicolaïdis 2021) and 'glocal' area studies (Kavalski in this volume; Roudometof 2015; Swynedouw 2004). Our contribution is three-fold: first, we refine the definition of resilience in International Relations; second, we scrutinize what elements facilitate resilience as transformation rather than simply adaptation; third, we zoom in on the local level of community and society, to bridge the research gap in IR studies between 'the global' and 'the local'. This Special Issue therefore connects and focuses on 'the local' to understand how resilience works in practice in times of complexity. The next section unpacks it through its constitutive elements.

What makes communities resilient: unpacking the fundamentals
Community resilience has been studied in different strands of literature, including disaster management (Imperiale and Vanclay 2016), ecology (Berkes and Ross 2013; Quinlan et al. 2016), psychology (Norris et al. 2008), anthropology (Barrios 2014; 2016; Tucker and Nelson 2017) and area studies (Anholt and Sinatti 2020; Petrova and Delcour 2020). Yet, to date the issue has remained a glaring blind spot in International Relations. By shifting the attention from the global to the local, resilience puts 'community' at the centre of analysis and engenders curiosity as to why some communities stay more resilient than others, even if they may have fewer resources and be less prosperous comparatively in material wealth. Drawing on the existing studies in community resilience, this section engages in an interdisciplinary conversation about the components of resilience to fill the knowledge gap about 'the local' in International Relations by offering a framework for analysis and understanding of community resilience, and its implications for different-level governance. However, before unpacking the fundamentals, two brief clarifications are in order: what is a community, and what does it mean to be resilient?

This edited volume refers to 'community' in a broad sense, as a group of individuals having a certain characteristic in common, including being bound, to a degree, by a specific locality, culture, behaviour, norms, institutions, and a 'shared vision'. 'Community' thus can refer to a family, neighbourhood area, districts, or civil society. Resilience, as discussed above, is understood both as a quality of a complex adaptive system with a range of components that make it enduring and responsive to change, and as an analytic of governance, a way of thinking and governing, that draws on self-reliance and self-organization, mobilising communities' inner strengths and capacities in the face of adversity, with external assistance as necessary. What follows below is an exploration and explanation of how resilience as a quality of a system may work in practice, through its multiple components, and what kind of governance-thinking it requires.

Identity and the meaning of a 'good life'

As Dooley notes, 'the desired state [of a complex adaptive system] is driven by and feedbacks to a "shared vision"' (1997, 91), critical to its survivability. In this subsection we will explore the role and the meaning of this *vision* for communal resilience-building as premised on the two important elements – *identity* and a sense of a *'good life'* – that glue communities together to make them resilient in the face of adversity. It is worth noting here that we treat 'identity' not as a stand-alone contributor to resilience, but as a *process* of making sense of, becoming with, and seeking a 'good life', which defines the human need for adaptation and change.

Much has already been said about identity, both temporally and across disciplines (Ohad and Bar-Tal 2009; Hall 1999; Katzenstein 1996; Neumann 1999; Wendt 1994). As Brubaker and Cooper (2000) note, identity has become an everyday idiom, being everywhere and nowhere at the same time, so much so that its vernacular overuse has led to an 'identity' crisis in social sciences. Without engaging with the vocabular utility of identity, in this Introduction we propose to link its epistemological meaning(s) to a new concept of a *'good life'* (Sadiki 2016; Flockhart 2020; Aristotle[7]). This concept is – akin to what Berenskoetter refers to as a 'future vision' (2011) – that stems from the uncertainty of the VUCA-world and individuals' desire for a more meaningful future and provides them with *possibilities of being* and *becoming with*, as a community, in the world.

In simple terms, identity is a human attempt, individual or collective, to 'establish a sense of Self in time' (Berenskoetter 2011, 648). Conventionally, it is understood as a social construct shaped by *the past* – that is, a shared understanding of history whereby 'actors see the future only through the strong filters of past socialization' (Copeland 2000:206), and embedded in *the present* – a shared culture, traditions, and norms 'as makers of (in)appropriate behaviour which are inscribed in routine practices and [upheld by] institutions located on the domestic or the international level' (Berenskoetter 2011, 650). However, what is often overlooked is the role of *the future and shared purpose*, in the human pursuit of survival and adaptation in a complex world.

Berenskoetter argues that, in an increasingly complex and unpredictable world, uncertainty plays a crucial role in identity formation: in particular, he notes that 'identity is [only] manifested through the future' where the latter is a 'source of anxiety' (citation). Identity 'renders being incomplete' (Berenskoetter 2011, 652). He draws his insights from Heidegger's work (1953), who insists that 'until it is dead there is always something the Self is not yet, and hence, "being" is always incomplete' (1953, in Berenskoetter 2011, 653). In the context of anxiety about the unknowable future, the identity of Self (singular or collective) is to a significant degree future-oriented, shaped by 'a desire to understand or give meaning' to the future (citation). Identity therefore 'renders the future *the most* significant parameter of being/becoming' (Berenskoetter 2011, 653). This meaningful future equates to a conception of a 'good life' defined predominantly in ideational rather than material terms – 'a sense of where we are going' (Anderson 2006), as an aspiration for 'a Significant We' (Flockhart 2006) to make 'the future Self "knowable"' (Berenskoetter 2011, 653), rational and worth living. This is a powerful drive not only for 'coping' with stress and adversity today, but also for seeking change and a better tomorrow, which lies at the heart of communal resilience-building. Heidegger refers to this driving force as *'Entwurf'*, which 'renders the future a "pull factor", providing the Self with an opportunity to move on, or ahead, on a certain [purposeful] course' (Berenskoetter 2011, 653). This sense of 'good life' lends the Self orientation and also the resolve and determination to realise a vision of becoming: 'one can argue that understanding and pursuing these possibilities, the Self already *is* these possibilities' (Berenskoetter 2011, 653, emphasis original). So, identity driven by a sense of 'good life' is a *process of becoming with*, which mobilises individuals with a shared purpose, to survive, adapt and transform together. This process, however, has two inherent *dualities*, the dynamics of which are contextually causal and important for understanding how local resilience comes about.

First, 'becoming' is always *intersubjective* in nature: it emerges as an intra- and inter-active mesh of Self and Other (Kurki 2020), in the process of their struggle and adaptation to internal and external environments. This duality of Self and Other is viewed differently in different traditions of thought and geographies of the world. If, for example, in liberal traditions the Self is seen as individual and central to defining relations, and is often situated either in opposition or juxtaposition to the (presumably inferior) Other (Diez 2005; Nicolaïdis et al. 2015), in several local traditions of Central Asia the concept of *barzak* indicates that the Self is always part of something bigger, more meaningful than its singular experience, something that even transcends death: *'You are everything, inside everything, and part of everything'* (Ibn Arabi cited in Nurulla-Khojaeva 2017, 119 and in this volume; see also Qin 2018; Green 2012). A reference to a 'mirror effect' is commonly used in Central Asia to transcribe this sense of collective being (see Nurulla-Khojaeva in this volume) or *becoming with*, which Chandler develops further in this issue: when one looks in the mirror, they do not see themselves but a world around them as *together-ness*. Hence, the importance of *'hamsoya'* (sharing a shadow with your neighbour) and *'suzami'* (a symbol of unity) that come to represent the primacy of a collective Other in a Self's becoming *with the community of beings*, informing a recurrent philosophy of resilience across Central Eurasia that makes a sense of community highly tangible.

Second, 'becoming' is a balance between stability and change. Identity is both affirmation of one's belonging to give some situational *certainty* (as part of the anxiety-controlling mechanisms) and it is a process of *change*. Identification is an assemblage of (i) the assumed desire for stability (what am I?) and (ii) the conception of the Self as evolutionary (always in the making), aiming to adapt and transform. As a process, Hopf (2002) argues, identity is about making the unfamiliar familiar (stabilization) and future visions more tangible (change). Therefore, 'becoming' is *a continuing process of identification* and transformation, in search for a 'good life', and equilibrium. Finally, identity of the Self and its 'future vision' of the 'good life' are two sides of the same coin – of the process of *becoming* when turning *irrational* reactions to change (our identity) into *rational* visions of the future, the construction of which is based on memories, experience, group socialization, resources, desires and dreams (Berenskoetter 2011).

The 'good life' thus is a possible utopia (or a vision of the future): a source of energy which motivates, mobilizes and moves the Self forward; it is perhaps the only rational thing in the arsenal of Self. Berenskoetter distinguishes between robust (certain/predetermined) and creative (able to open political spaces) visions. Based on Berenskoetter's analysis, and the empirical contributions to this volume, we suggest that Central Eurasia provide more fertile ground for *creative visions* to emerge. Such visions are driven by an idea(l) that connects past philosophies of life with future aspirations, and creates *a sense of becoming with*, which promises to transform the established order of things. For a vision to be attractive for sharing/following, it must resonate with shared cultures, philosophies, and traditions, while also offering an alluring 'promised land' of hope and goodness. When faced with adversity, these visions of a better tomorrow would stimulate the mobilization of inherent resources, communal support infrastructures and the resolve (grit/tenacity/strength) needed to cope with crisis.[8] A sense of the 'good life' needs to function as a creative (ideational) space and accommodate various forms of evolving multiple interpretations (through dialogue[9]), the blueprint of which is always typically local, indigenous.

Inherent resources and community support infrastructures

Identity and the pluriversal vision of a 'good life' are the driving force for communal adaptation and transformation in search of a better tomorrow. Yet, in an everyday life riddled with uncertainties and irrationalities, communities also require some more tangible forms of support – as defined by their networks of relations (communal infrastructures) and resources – to help them survive and adapt. Once again, Central Eurasia, like some other 'developing' localities, often presents communal support infrastructures that distinctively rely on significantly informal and dense relations of responsibilities (from moral to financial) as well as a stronger collective safety net for supporting the vulnerable and the needy (see e.g. Hutchinson and Korosteleva 2006; Badescu and Uslaner 2003 and empirical contributions to this issue). In this subsection we will explore what tangibly makes communities more than just a gathering of persons by zooming i on the formal and informal community structures and resources.

Complexity- and resilience-thinking is based on the notion of emergence, also referred to as *self-organization*. Emergence can be defined as the interaction of individual units, without governance or coordination from above, that results in an outcome qualitatively different from the aggregate of individual inputs. Passing through feedback loops, these outcomes may evolve into *orders*, facilitating resilience of a system/community. An order, as defined by Lebow (2018, 8), is 'a hierarchical arrangement, supported by most of its members, that fosters security, self-esteem, and social contract, encourages solidarity, and results in legible, predictable behaviour'. Given the non-linearity and processual nature of emergence, orders and their constituent elements are deeply embedded in spatio-temporal contexts (see Flockhart in this issue). This implies that ultimately there are no universal solutions. Emerging structures for coping that may come to constitute orders are highly context-specific, which explains the exuberant mosaic (or 'pluriverse') of community resilience strategies (Kothari et al. 2019).

Some common elements of self-organization identified in community resilience literature include economic elements (equity of resource distribution, diversity of economic resources), information and communication (narratives, trusted sources of information), social capital (social ties and networks, citizen participation, leadership, trust, reciprocity, attachment to place, etc.) and community competences (community action, empowerment, sense of community) (Norris et al. 2008; Berkes and Ross 2013). Recent literature showcased prominent examples of self-organization reflecting the importance of these elements. For instance, *agaciro* (dignity, self-worth), a philosophy and policy originating from Rwanda, aims to move away from dependence and international aid, and replace them with self-reliance and solidarity (Rutazibwa 2014). It puts forward the vision of relationality and self-help, emphasising local structures as more attuned to people's aspirations as compared to global development discourses. *Agaciro* is echoed in the Andean concept of *buen vivir* (good living), explained in a nutshell as 'collective well-being according to culturally appropriate conceptions' (Escobar 2018, 148). *Buen vivir* is an empowering vision acknowledging multiple development paths, plurality of local knowledges, and relational understanding of life. The emphasis on relationality and functioning community networks is emerging as the key to community resilience, as demonstrated by multiple empirical studies, including the contributions to this issue.

A more tangible fabric to help structure local communities and their order is embodied in a range of formal and informal support infrastructures. Formal or institutionalized forms of bringing order into communal living take multiple variations in Central Eurasia, e.g., *mahalla*, community of elders in Central Asia, *tovarischestva* in Russia or *supol'nast'* in Belarus. While not initiated by the state, these societal forms of self-organisation are defined as relatively formal due to their institutionalization, manifested via assigned roles, inherent hierarchies (*stareishyna* or *aksakal* in Central Asia), possession of community funds, legitimacy and authority. Instead, informal community infrastructures are more fluid. They include family, kinship, friends and neighbour networks, as well as occasional traditional gatherings (festivals, weddings, funerals etc.) and dedicated community support groups in case of an emergency or specific event/need. For instance, the Slavic, and particularly Belarusian, term *talaka* (self-help) historically referred to a gathering of a neighbourhood to complete some labour-intensive work together, such as harvest-gathering or house-building. With the development of digital technologies, informal infrastructures have gained traction in the forms of neighbourhood chats and online communities, boosting societal ties and facilitating local cooperation among community members. This was strongly evidenced by communal crowdfunding, and various support measures for the vulnerable during the COVID- 19 pandemic. This variety of community networks provides community services, but also essentially serves as a source of ontological security, necessary for resilient adaptation to a changing environment, which will be exemplified further by the empirical part of this volume.

Peoplehood as mobilization of resilient communities

In the previous two subsections, we briefly unpacked the notions of identity ('what we are') linked to a 'good life' ('what we want to be') in *the process of becoming with*. We also illustrated how local formal and informal support infrastructures could help communities stay stronger *together* through *self-organisation* and *emerging order*, when facing the challenges of uncertain future, and threats to their internal and external environment. Drawing on the empirical analysis of our case studies, as well as theoretical discussions of philosophy, locality, poetics and the opacity of human relations presented in this volume, we believe these are encompassing but not necessarily exhaustive components of building and maintaining *resilience*, which help communities survive adversity and transform under the pressure of change.

In this subsection we introduce one more component of resilience – *the people-hood* – which is not commonplace, but which signifies *the moment of becoming with*, when all resources, capacities and future 'visions' that give a community of relations a more consolidated quality align with each other to take it to a new level of *being together*. The peoplehood is often mobilized at the moment of existential threat and severe violation of a community's fledgling foundations. This mobilisation was famously captured, for example, by the Arab Spring, described as '*al-harak*', that is, 'the essence of the political, social, cultural, and religious people-driven ferment' (Sadiki 2016, 339); or, by the moment of 'the revolution of Dignity' turning ancient monasteries into battlefield hospitals in Kyiv (from 2014 onwards); or by the defiant and pervasive resistance in Belarus post-presidential election in 2020, not submitting to the oppression of the regime. In these instances, people reached *the moment of becoming with*, a qualitatively different political entity, with a sense of dignity (*agaciro*) and self-worth to fight for and protect their future.

'Peoplehood' as *becoming* and *being with* is deeply transformative and vehemently powerful (Korosteleva and Petrova 2021). It is also *political* – seeking to transform the environment, rather than adapt to survive. This is a relatively new concept in social sciences, which is yet to develop a unified and clear meaning. It has emerged with the intensifying levels of people's engagement in politics, driven by a strong desire to make their lives more equitable, fair and sustainable. Smith (2015, 3), for example, contends that peoplehood is more than just becoming 'political people': it is about 'conveying senses of meaning and value, defining political goals, prescribing institutions and policies, and sustaining or failing to sustain support for political communities and their leaders, institutions and policies in difficult times'. According to Lie (2004), peoplehood offers an inclusionary and even involuntary group identity with a putatively shared history and distinct way of life. He clarifies further: 'It is *inclusionary* because everyone in the group, regardless of status, gender, or moral worth, belongs. It is *involuntary* because one is born into an ascriptive category of peoplehood... It is not merely a population, but rather a people – a group, an internal conviction, a self-reflective identity' (2004, 1). Peoplehood, as a *moment of becoming*, acquires its own distinct discourse and a special identity of '*being together*, not merely in similar ways' (Brown and Kuling 1997, 43) challenging the status-quo, reinforced through the symbols of otherness (e.g., the white-red-white flag in Belarus), or an acute sense of injustice that may threaten survival (e.g., the 'Black Lives Matter' movement). Peoplehood becomes more than society (Dominiquez 1989): it turns into a transformative political entity, which comes to encapsulate fragile social relations and an urgent need to 'interact in ways other than through force or imposition' (Anderson 2014, 19). Sadiki (2016, 339) notes that the rise of peoplehood is an 'important watershed' in the life of society: 'it partakes of both civil and uncivil manifestations of thought and practice across boundaries of rich diversity and complexity', potentially even 'morphing into a transnational phenomenon' (Sadiki 2016, 339).

In conclusion, *resilience* as a *quality of a complex adaptive system* and a way of thinking begins at the local level, and is manifested via an assemblage of its constitutive elements, including (but not limited to) *identity* driven by a sense of a '*good life*' and its inherent duality, an awoken sense of self-worth and *dignity* (*agaciro*), and formal and informal communal *support infrastructures*, which could turn existing capacities into true *capabilities of peoplehood* to fight for a better future when faced with an existential threat. This non-exhaustive list of resilience components underscores the primacy of '*the local*' and its potential to make global governance in a more complex and uncontrollable VUCA-world more sustainable and effective through self-organisation and *self-governance*.

Why Central Eurasia: exploring its internal and external dimensions

Having unpacked the fundamentals of resilience as a quality of a complex system, and as a process of self-governance, we shall now briefly explain why Central Eurasia was singled out for understanding the workings of resilience in a complex world. As Scott Levi stated, 'what we are dealing with are not separate and comparable, but connected histories... [These networks] were the avenues through which knowledge of the outside world reached Central Asia, and they were *extraordinarily resilient*' (2020:170; emphasis added). In other words, while we see Central Eurasia as a particularly illustrative locality for the purpose of this research, it should not be understood monolithically in isolation from its global environment. Rather, it preserves transnational networks that are historic-ally rooted and make its experience significant in rethinking resilience on a global scale. We therefore see Central Eurasia as a powerful locus for a new study of resilience in IR, both historically and in contemporary world politics.

There are three essential reasons to justify our choice of locality. First, Central Eurasia is considered a rich unfolding universe shaped by a centuries-long his-tory of global connectedness, remarkable fluidity as an inherently nomadic space (Levi 2020; Hansen 2012; Frankopan 2015), and endurance as a way to adapt and transform, thus underscoring its inherent resilience. It has been defined as a 'crossroads of civilisations' (Folz 2016) whose mission – Abu Hamid Al-Ghazali believed – was 'to connect and resolve the controversy between the worlds and the human' via the ideas of *dahleez* (a door between the worlds) and *barzak* (Nurulla-Khojaeva 2017, 122, and in this volume). Several ancient and modern scholars point to Central Eurasia as a mesmerizing cradle of 'lost' wisdom and newly-found enlightenment, the homeland of thinkers who 'affected science and civ-ilization' globally, connecting 'antiquity and the modern world' (Starr 2013: 4, 21), as well as a locus of extraordinary skills, knowledge and cultural diplomacy, epitomised by the Sogdian merchants who populated Central Eurasia from the VI century BC to the XVI AC (Nurulla-Khojaeva 2017). Remarkably, the philosophy of Sufism, which has permeated Central Eurasia for centuries through artistic and poetic production (Green 2012), still arguably remains a strong 'pull factor'[10] for survival and transformation to this day (see. e.g. the work of Nasritdinov and O'Connor 2009; Peyrouse and Nasritdinov 2021). This extraordinary space has also been the focus of the COMPASS research project, embracing Belarus in the west, Azerbaijan in the south and Tajikistan in the east, which made this pioneering research into the resilience of Central Eurasia possible.

The second reason relates to the very nature of the communities that charac-terise Central Eurasia as a locality, making it insightful for both historical and con-temporary study of resilience (see for e.g Neumann and Wigen 2018; Reynolds 2020; White 2020). Central Eurasia is home to peoples who lived through centuries-long hardship and depravity, and yet saw beauty and poetics in every-thing and learned to adapt, share and transform in their processes of becoming and reaching toward their visions of happiness, good life, and good neighbour-liness. And yet, Central Eurasia is extraordinarily understudied compared to the body of scholarship exploring societies in Latin America and Africa, their efforts at decolonization and post-development, as well as indigenous ideologies for progression and visions of the future, as shown by Kalra in this issue. Central Eurasia is fraught with massive challenges, including ongoing transitions, limited resources, rampaging poverty, lasting conflicts, health and environmental crises, as vividly demonstrated by Babaev and Abushov as well as Markovich et al. in this issue. Yet, its peoples still survive and prosper, thus providing a remarkable case-study for understanding what makes communities so resilient there, and what their governance-thinking could teach us, especially in these troubling days of health emergencies, environmental calamities, and economic and political crises.

Finally, Central Eurasia is also remarkable in terms of its geo-strategic location: it spans two continents, and is at the epicentre of interest and investment from at least three major global powers – the European Union (EU), Russia and China – each projecting their own visions and governing strategies to engender, as they claim, growth, prosperity and stability there.[11] And yet, all three powers often assume too much knowledge and understanding of this diverse and polyphonic locality, this way lessening its own agency, sustainability and self-governing opportunities (Kavalski 2012; Korosteleva and Petrova 2020; Kalra; Bossuyt and Davletova in this volume). Understood heuristically as a locality which is distinctive but diverse, Central Eurasia thus troubles familiar conceptions of the international and world politics in IR, too often understood from the totalising perspectives of great powers or seemingly uniform geopolitical wholes (e.g., "the West" and "the non-West"). We hope this study of resilience, order and governance of *the local* will alter and unsettle these trajectories of learning by placing Central Eurasia as a driving force of resilient development firmly on the study map of International Relations.

The Edited Volume's structure and contributions

This edited volume makes a substantive contribution, in theory and practice, to the study of International Relations, by focusing on resilience's constitutive elements to understand (1) what helps communities survive, adapt and transform; (2) how orders form; and (3) what kind of governance is needed to make 'the global' more sustainable through 'the local' in times of growing complexity and change.

The volume offers theoretical, conceptual and empirical perspectives on the study of societal resilience and its core components. After an introduction that outlines the overarching framework of its relational elements, the discussion first moves to consider alternative *framings of resilience* as poetics of relations to be decolonized from 'Western' (neoliberal) narratives (Chapter 1 by David Chandler), and theorise *the role of a 'good life'* in shaping a resilient order in a multi-order world (Chapter 2 by Trine Flockhart) and rethinking of International Relations altogether from a *complexity-thinking perspective* (Chapter 3 by Emilian Kavalski). Most notably, Chandler argues that understanding resilience means allowing the *opacity* of processual *becoming with others in relation* to take its course through improvisation and feedback loops, which in turn would push communities to experiment, to be creative, and to draw on their inherent capacities and visions of the future to change as a collective. Crucial here, Chandler asserts, is the *conception of relation*, as explored by Glissant (1997) when postulating 'the right to opacity' and further developed into new resilience approaches by Kara Keeling (2019) and An Yountae (2017). The notion of relation and its opacity, as Chandler contends, 'keep communities open to changes which cannot be predicted beforehand', and in this way allows them to 'grow and develop as they "world themselves" in an open set of responsivities', rather than via closed choices, enforced solutions or fixed identities. Being open in and to the world always places one 'in the middle of *processes of inter-relation*', thus not only engendering diversity, but also encouraging curiosity for 'alternative futural imaginaries' while continually constructing a community of relations.

A community of relations is bound together by a sense of a *'good life'*, as Trine Flockhart argues in her piece. In particular, while 'order is a fundamental condition for social life', she further contends that what keeps social life together is a *shared vision* and values that constitute the aspirational notion of a 'good life'. This in turn raises some crucial questions of whose vision for a "good life", and whose order will count, which are fundamental for the resilience of international order(s). Flockhart insightfully examines a crisis of the international liberal order, as a 'local process' invariably connected to and in turn impacting the global architecture. She questions how global international society can become more sustainable and how competing visions for a 'good life' can co-exist. Ultimately, she concludes, what matters for making 'the global' more responsive in a *complex multi-order world* is a diversity of being, which propels the need for dialogue with 'the local'. The discussion takes a new turn in Chapter 3 by Emilian Kavalski who challenges the prevalent 'castle-thinking' approach among the IR scholars who invent and defend the 'bulwarks of their own analytical castles' as stake-claiming rather than explaining the increasingly complex world around us. He proposes to rethink, and reinvigorate IR by way of complexifying it, to see it as a world of relations, at the heart of which is a community of human and non-human subjects, and which actions and interactions are so interwoven that they define the global dynamics of the increasingly 'local world' of relations.

These theoretical discussions are followed by philosophical and conceptual explorations with a focus on Central Eurasia. In particular, Nargis Nurulla-Khojaeva (Chapter 4) examines the meaning of resilience in Sufi traditions by delving into Central Asian philosophical, religious and poetic foundations. She posits that their thinking is intimately interwoven with local traditions of 'listening with the ears of the heart' and the world, through a centuries-long connectedness and *remembering*. At the heart of it all is the notion of *genuine resilience (hamsoya)*, as the art of listening and feeling your neighbour, being part of one 'whole' (*hamdili*). There is never a singular 'I', but always a shadow of 'We' that provides comfort, dignity, care and a measure of the world around us, like the 'seven birds of *hamsoya*' representing polyphonic voices of the region and many meanings of the world. What matters here is how 'the local' always stays connected, and how genuine resilience of communities, while opaque and hidden, makes every person an intrinsic part of the global world. This is followed by the conceptual discussions of communal self-governance as an alternative to neo-liberal thinking of the European Union, developed by Fabienne Bossuyt and Nazima Davletova in Chapter 5 of this volume. There they claim that in order for the EU governance to become more effective and sustainable, it needs to decentre and accept 'the other' – in this case, the Central Asia societies – for what they are, with their traditions of neighbourliness and *hamsoya*, based on a deep understanding of the local meaning of the good life, and local knowledge about the available resources.

They develop their investigation by using first-hand evidence in their study of Uzbekistan and the wider region.

The volume proceeds further by offering empirical explorations of Central Eurasia, to show how 'the local' always stays connected with its past and the future, and how resilience of communities, while opaque and hidden, makes every person an intrinsic part of the global world. Hence, it is of critical importance to study *communal relations* and their resilience, especially, as Chandler concedes, through poetics, which render immense energy of *becoming* into the world. The articles explore what communal resilience means in practice by looking at communities across Central Eurasia. Belarus in particular, as examined by Anna Markovich et al. (Chapter 6), presents an insightful case of community, of *becoming with others in relation*, whereby a centuries-long endurance and a sense of a 'good life' have been momentarily transformed into *peoplehood* in response to injustice and lack of Covid-related state care. It is incredibly powerful to observe a palpable mushrooming of hitherto fragmented communal gatherings (*supol'nast'*), which emerge as self-organization to support, care for and protect each other. A similar wave of transformation, as analysed by Azer Babaev and Kavus Abushov in Chapter 7, is noticeable in Azerbaijan, recently hit by a Karabakh war, exposing the process of *peoplehood-in-the-making* through *affective solidarity* and a surge of communal support infrastructures through kinship and neighbourhood ties. In Chapter 8 Chiara Pierobon and Zarina Adambussinova explore societal resilience in Kyrgyzstan drawing on original fist-hand data, to understand what makes some local communities stronger than others, and what could enable more sustainable interventions to support them further in their development, without stifling their indigenous visions of the good life, and aspirations for the future.

The final contribution by Prajakti Kalra offers an exciting account of *historical developments* in Central Eurasia across the centuries of trade, culture and nomadic mobility of what has long been known as the Silk Roads, inextricably linked with human resilience. As Kalra argues, it is this inherent resilience, 'hybridity, trans- and multi-culturalism', along with its abiding history and local polyphony, that make Central Eurasia so enchanting and important to give heed to if one wishes to develop a better understanding of complexity and relationality in the world we live in today, and to learn to make governance more sustainable. This chapter concludes the discussion of resilience in Central Eurasia by emphasising its inherited strength, and its genuine organic embeddedness into the philosophy, culture and daily living of people across this vast and incredibly diverse locality.

Acknowledgements
We would like to thank the ESRC funders and Research England fund at Warwick University for supporting our research in the region; all contributors to this Volume for their insightful articles; and the anonymous reviewers, the CRIA and SPIB series editors for their support.

Notes

1. This part of the definition draws extensively on the works of Bourbeau 2018; Krause 2018; Joseph 2018 and many more.
2. We see our work on resilience as an analytic of governance essentially as a three-step inquiry. First, we explored the notion of resilience as a nexus between 'the global' and 'the local', to define it as a self-governing system of local communities (Korosteleva and Flockhart 2020). In this volume, as a second step, we unpack its fundamentals to understand how resilience works in practice. The third step will be to connect 'the local' back to 'the global', in search for more cooperative ways of all-level governance in a complex world, where 'many worlds fit' (Escobar 2018). The latter, while important, is outside the scope of this inquiry, although some arguments in this volume (e.g. Flockhart; Chandler) already allude to how it could be done.
3. For more information see: https://research.kent.ac.uk/gcrf-compass/
4. The concept of 'the mesh' was developed by Morton (2010; 2013) to account for the totality of relations and relationalities of the world and the universe. For the concept of the mesh in International Relations, see Kurki (2020).
5. See Bousquet and Geyer (eds.) (2011) 'Complexity and the international arena', *Cambridge Review of International Affairs*, 24:1 and Nordin et al (2009) 'Towards global relational theorizing', *Cambridge Review of International Affairs*, 32:5.
6. For detailed explanation of complexity-thinking, particularly in the context of international affairs, see Bousquet and Geyer (eds.) (2011) 'Complexity and the international arena', *Cambridge Review of International Affairs*, 24:1.
7. Aristotle's discussion of happiness, good and goodness, could be revisited in his own work, especially on Nicomachean Ethics (*Aristotle's Nicomachean Ethics*, Robert C. Bartlett, and Susan D. Collins (eds/trans.), Chicago: The University of Chicago Press, 2012); and Eudemian Ethics (*Eudemian Ethics*, Cambridge Texts in the History of Philosophy, Brad Inwood and Raphael Woolf (eds./trans.), Cambridge: Cambridge University Press 2013).
8. Belarus post-election 2020 is a good point of reference: a newly mobilized identity of being/becoming Belarusian, associated with anti-violence, and national symbols, self-mobilized itself, by connecting to the past and striving for a peaceful and democratic vision of tomorrow.
9. See Reus-Smit 2018 for further thinking on the relevance of cultural diversity for resilient order.
10. See also this interview with Sebastien Peyrouse about religion in Central Asia available here: https://cabar.asia/en/what-is-the-situation-with-religious-education-in-cent ral-asia-interview-with-sebastien-peyrouse
11. For more discussion see a Special Issue by Korosteleva, E. and Paikin, Z. (2021) 'Russia between east and west, and the future of Eurasian order', *International Politics* 58:3.

Disclosure statement

No potential conflict of interest was reported by the authors.

Funding

This work was supported by the GCRF-funded COMPASS project 'Comprehensive Capacity-Building in the Eastern Neighbourhood and Central Asia: research integration, impact governance and sustainable communities' (ES/P010849/1), and Participatory Research England Fund, University of Warwick.

References:

Acharya, A and B Buzan (2017). 'Why is there no non-Western international relations theory? Ten years on'. *International Relations of the Asia-Pacific*, 17:3, 341–370, https://doi.org/10.1093/irap/lcx006.

Anderson, C (2014) *Metis: race, recognition, and the struggle for indigenous people* (UBC Press).

Anholt, R and G Sinatti (2020) 'Under the guise of resilience: the EU approach to migration and forced displacement in Jordan and Lebanon', *Contemporary Security Policy*, 41:2, 311–335, DOI: 10.1080/13523260.2019.1698182

Babaev, A and K Abushov (2021) 'Azerbaijanis' resilient society: Explaining the multifaceted aspects of people's social solidarity', *Cambridge Review of International Affairs* (and in this volume).

Badescu, G and E Uslaner (2003, eds.) *Social capital and the transition to democracy* (Routledge).

Bargués-Pedreny, P (2020) 'Resilience is "always more" than our practices: limits, critiques, and scepticism about international intervention', *Contemporary Security Policy*, 41:2, 263–286, DOI: 10.1080/13523260.2019.1678856

Barrios, R (2014) "Here, I'm not at ease': Anthropological perspectives on community resilience', *Disasters*, 38:2, 329–350.

Barrios, R (2016) 'Resilience: A commentary from the vantage point of anthropology', *Annals of Anthropological Practice*, 40:1, 28–38.

Berkes, F and H Ross (2013) 'Community resilience: toward an integrated approach', *Society & Natural Resources*, 26:1, 5–20, DOI: 10.1080/08941920.2012.736605

Berenskoetter, F (2011) 'Reclaiming the Vision Thing: Constructivists as Students of the Future', *International Studies Quarterly*, 55:3, 647–668, doi.org/10.1111/j.1468-2478.2011.00669.x

Bossuyt, F., and Davletova, N. (2022) 'Communal self-governance as an alternative to neoliberal governance: proposing a post-development approach to EU resilience-building in Central Asia', *Central Asian Survey* (and in this volume).

Bourbeau, P (2018) *On resilience: Genealogy, logics and world politics* (Cambridge University Press).

Bousquet, A and R Geyer (2011) 'Introduction: Complexity and the international arena', *Cambridge Review of International Affairs*, 24:1, 1–3, DOI: 10.1080/09557571.2011.558713

Bousquet A and S Curtis (2011) 'Beyond models and metaphors: Complexity theory, systems thinking and international relations', *Cambridge Review of International Affairs*, 24:01, 43–62, DOI: 10.1080/09557571.2011.558054

Brown, D and J Kulig (1996/97) 'The concept of resiliency: theoretical lessons from community research', *Health and Canadian Society*, 4, 29–52, p. 43

Brubaker, R and F Cooper (2000) 'Beyond 'identity'', *Theory and Society*, 29:1, 1–47, https://doi.org/10.1023/A:1007068714468

Burrows, M and O Gnad (2017) 'Between 'muddling through' and 'grand design': Regaining political initiative—the role of strategic foresight', *Futures* 97, 6–17.

Chandler, D (2014) 'Beyond neoliberalism: Resilience, the new art of governing complexity', *Resilience*, 2:1, 47–63, DOI: 10.1080/21693293.2013.878544

Chandler, D (2020) 'Security through societal resilience: Contemporary challenges in the Anthropocene', *Contemporary Security Policy*, 41:2, 195–214.

Chandler, D (2018) *Ontopolitics in the Anthropocene: an Introduction to Mapping, Sensing and Hacking* (Routledge).

Chandler, D (2022) 'Becoming resilient otherwise: Decolonising resilience approaches via Glissant's poetics of relation', *Cambridge Review of International Affairs* (and in this volume).

Chandler, D, Grove, K and S Wakefield (2020, eds.) *Resilience in the Anthropocene: Governance and politics at the end of the world* (Routledge).

Chandler, D and J. Reid (2016) *The Neoliberal Subject: Resilience, Adaptation and Vulnerability* (Rowman & Littlefield).

Chandler, D, Müller, F and D Rothe (2021, eds.) *International relations in the Anthropocene: New agendas, new agencies and new approaches* (Palgrave Macmillan).

Clark N and B Szerszynski (2021) *Planetary social thought: The Anthropocene challenge to the social sciences* (Polity Press).

Copeland, D (2000) 'The constructivist challenge to structural realism', *International Security* 25:2, 187–212.

Cusumano, E and S Hofmaier (2020) *Projecting resilience across the Mediterranean* (Palgrave Macmillan).

Edkins, J (2019) *Change and the politics of certainty* (Manchester University Press).

Diez, T (2005) 'Constructing the Self and changing Others: Reconsidering 'Normative Power Europe'', *Millennium: Journal of International Studies*, 33:3, 613–636.

Dominquez, VR (1989) *People as subject, people as object: Selfhood and peoplehood in contemporary Israel* (The University of Wisconsin Press).

Dooley, K (1997) 'A complex adaptive systems model of organization change', *Nonlinear Dynamics, Psychology, and Life Sciences*, 1:1, 69–97.

Eoyang, G and T Berkas (1998) '*Evaluation in a complex adaptive system*', in Lissak, M and H Gunz (eds.) *Managing complexity of organisations: A view in many directions* (Praeger).

Escobar, A (2018) *Designs for the pluriverse: Radical interdependence, autonomy, and the making of worlds* (Duke University Press).

Fisher Onar, N and K Nicolaïdis (2021) 'The decentring agenda: A post-colonial approach to EU external action' in: Gstöhl, S and S Schunz (eds.) *Studying the European Union's external action: concepts, approaches, theories* (Macmillan International).

Finkenbusch, P. (2020) 'Beyond liberal governance? Resilience as a field of transition', *Journal of International Relations and Development*, https://doi.org/10.1057/s41 268-021-00207-1

Flockhart, T (2006) ''Complex socialization': A framework for the study of state socialization', *European Journal of International Relations*, 12:1, 89–118.

Flockhart, T (2016) 'The coming multi-order world', *Contemporary Security Policy* 37:1, 3–30.

Flockhart, T (2020) 'Is this the end? Resilience, ontological security, and the crisis of the liberal international order', *Contemporary Security Policy*, 41:2, 215–240.

Flockhart, T (2022) 'The Liberal International Order in transformation: Whose vision for the 'good life' will matter?', *Cambridge Review of International Affairs* (and in this volume).

Foltz, R (1999) 'The role of the Sogdians in the spread of world religions', Papers in Honour of Professor Z. Zarshenas. Available at www.cais-soas.com/CAIS/Religions/iran ian/role_central_asian_spread_religion.htm

Frankopan, P (2015) *The Silk Roads: a new history of the world* (Bloomsbury Publishing).

Green, N (2012) *Sufism: A global history* (Wiley-Blackwell).

Gell-Mann, M (1995) *The quark and the jaguar: Adventures on the simple and the complex* (London: Abacus).

Glissant, E (1997) *Poetics of relation* (University of Michigan Press).

Joseph, J (2013) 'Resilience as Embedded Neoliberalism: A Governmentality Approach', *Resilience*, 1:1, 38–52.

Joseph, J (2018) *Varieties of Resilience: Studies in Governmentality* (Cambridge University Press).

Jørgensen, S. E. (1990) Ecosystem theory, ecological buffer capacity, uncertainty and complexity. *Ecological Modelling*, 52, 125–133.

Hall, R (1999) *National collective identity: Social constructs and international systems* (New York: Columbia University Press).

Hansen, V (2012) *The Silk Road: A new history* (Oxford University Press).

Holland, J (1995) *Hidden order: How adaptation builds complexity* (Reading, MA: Addison Wesley).

Hopf, T (2002) *Social construction of international politics: Identities and foreign policies, Moscow, 1955 & 1999* (Ithaca, NY: Cornell University Press).

Hopf, T and B Bentley (2016, eds.) *Making identity count: Building a national identity database* (New York, NY: Oxford University Press).

Hutchinson, D and E Korosteleva (2006, eds.) *The quality of democracy in post-communist Europe* (Routledge).

Imperiale, A and V Frank (2016) 'Experiencing local community resilience in action: Learning from post-disaster communities', *Journal of Rural Studies*, 47, 204–219, doi.org/10.1016/j.jrurstud.2016.08.002

Juncos, A (2017) 'Resilience as the new EU foreign policy paradigm: a pragmatist turn?', *European Security*, 26:1, 1–18, DOI: 10.1080/09662839.2016.1247809

Kalra, P (2021) 'Resilient histories: Eurasia's moment and method to regain its historical legacy', *Cambridge Review of International Affairs* (part of the special Issue).

Katzenstein, P (1996, ed.) *The culture of national security: norms and identity in world politics* (New York: Columbia University Press).

Kauffman, S (1995) *At home in the universe: The search for laws of self-organization and complexity* (Oxford University Press).

Kavalski, E (2007) 'The fifth debate and the emergence of complex international relations theory', *Cambridge Review of International Affairs* 20:3, 433–454, https://doi.org/10.1080/09557570701574154

Kavalski, E (2012) 'Waking IR up from its 'deep Newtonian slumber'', *Millennium: Journal of International Studies*, 41:1, 137–150, https://doi.org/10.1177/0305829812451717

Keukeleire, S, Lecocq, S and F Volpi (2020) 'Decentring norms in EU relations with the Southern Neighbourhood', *Journal of Common Market Studies*, Ahead of print publication

Keeling, K (2019) *Queer times, black futures* (NYU Press).

Korosteleva, E (2018) 'Paradigmatic or critical? Resilience as a new turn in EU governance for the neighbourhood', *Journal of International Relations and Development*, 23, 682–700.

Korosteleva, E and T Flockhart (2020) 'Resilience in EU and international institutions: Redefining local ownership in a new global governance agenda', *Contemporary Security Policy*, 41:2, 153–175.

Korosteleva, E. and I Petrova (2021) 'Community resilience in Belarus and the EU response', *Journal of Common Market Studies*. Annual Review, Early view available at: https://onlinelibrary.wiley.com/doi/10.1111/jcms.13248

Korosteleva, E and Z Paikin (2021) 'Russia between east and west, and the future of Eurasian order', *International Politics*, 58:1, 321–34.

Kothari, A, Salleh, A, Escobar, A, Demaria, F and A Acosta (2019, eds.) *Pluriverse: A post-development dictionary* (Tulika Books).

Krause, J (2018) *Resilient communities: Non-violence and civil agency in communal war.* (Cambridge University Press).

Kurki, M (2020) *International relations in a relational universe* (Oxford University Press).

Latour, B (2017) *Facing Gaia: Eight lectures on the new climatic regime* (Cambridge: Polity).

Lebow, N (2018) *The rise and fall of political order* (Cambridge University Press).

Levi, S (2020) *The Bukharan crisis: A connected history of 18th-century Central Asia* (University of Pittsburgh Press).

Lie, J (2004) *Modern peoplehood: On race, racism, nationalism, ethnicity and identity* (University of California Press).

Luhmann, N (1990) *Essays on self-reference* (New York: Columbia University Press).

Macy, J (2007) *World as lover, world as self: Courage for global justice and ecological renewal* (Berkeley, CA: Parallax).

Manson, S (2001) 'Simplifying complexity: A review of complexity theory', *Geoforum* 32:3, 405–414.

Morton, T (2010) *The ecological thought* (Boston: Harvard University Press).

Morton, T (2013) *Hyperobjects: Philosophy and ecology after the end of the world* (London: University of Minnesota Press).

Mishra, P (2020) 'Grand illusions', *The New York Review*. Available at: www.nybooks.com/articles/2020/11/19/liberalism-grand-illusions/

Nasritdinov, E and K O'Connor (2009) *Regional change in Kyrgyzstan: Bazaars, cross-border trade and social networks* (Saarbrücken, Lambert Academic Publishing).

Norris, F, Stevens, SP, Pfefferbaum, B, Wyche, K and R Pfefferbaum (2008) 'Community resilience as a metaphor, theory, set of capacities, and strategy for disaster readiness', *American Journal of Community Psychology*, 41:1–2, 127–150, DOI 10.1007/s10464-007-9156-6

Neumann, I (1999) *Uses of the Other: The East in European identity formation* (Minneapolis: University of Minnesota Press).

Neumann, I and E Wigan (2018) *The Steppe Tradition in International Relations*. Cambridge University Press.

Nicolaidis, K et al. (2006, eds.) *Echoes of empire: Memory, identity, and colonial legacies* (London: Tauris).

Nordin, A, Smith, G, Bunskoek, R, Hwang, C, Thaddeus Jackson, P, Kavalski, E, Ling L. H. M, Leigh Martindale, et al. (2019) 'Towards global relational theorizing: a dialogue between Sinophone and Anglophone scholarship on relationalism', *Cambridge Review of International Affairs*, 32:5, 570–581, DOI: 10.1080/09557571.2019.1643978

Nurulla-Khojaeva, N (2017) "Dancing' merchants beyond the empires of the Silk Road', *Vestnik MGIMO*, 1:52, 119–39 [in Russian].

Nurulla-Khojaeva, N (2022) '"Imitated" or genuine? The value of resilience in Sufi-hamsoya' (in this volume)

Ohad, D and D Bar-Tal (2009) 'A sociopsychological conception of collective identity: The case of national identity as an example', *Personality and Social Psychology Review*, 13:4, 354–379.

Petrova, I and L Delcour (2020) 'From principle to practice? The resilience–local ownership nexus in the EU Eastern Partnership policy', *Contemporary Security Policy*, 41:2, 336–360.

Pierobon, C. and Adambussinova, Z. (2022) 'Community resilience of post-Soviet mono-industrial areas affected by the uranium legacy and radiation: evidence from Kyrgyzstan' (in this volume).

Pravdivets, V., Markovich, A., and A. Nazaranka (2021) 'Belarus between West and East: experience of social integration via inclusive resilience', *Cambridge Review of International Affairs* (and in this volume).

Qin, Y. (2018) A Relational Theory of World Politics (Cambridge University Press).

Quinlan, A, Berbes-Blasquez, M, Haider, J and G Peterson (2016) 'Measuring and assessing resilience: Broadening understanding through multiple disciplinary perspectives', *Journal of Applied Ecology*, 53:3, 677–687.

Rouet, G and G Pascariu (2019) *Resilience and the EU's Eastern Neighbourhood countries: From theoretical concepts to a normative agenda* (Palgrave Macmillan).

Reynolds, M (2020) An original and thought-provoking first crack at the Steppe in IR, *Cambridge Review of International Affairs*, 33:6, 931–936, DOI: 10.1080/09557571.2020.1838206

Reus-Smit, C (2018) *On cultural diversity: International theory in a world of difference* (Cambridge University Press).

Roudometof, V (2015) 'The Glocal and Global Studies', *Globalizations*, 12:5, 774–787, DOI: 10.1080/14747731.2015.1016293

Rutazibwa, O (2014) 'Studying agaciro: moving beyond Wilsonian interventionist knowledge production on Rwanda', *Journal of Intervention and State-building*, 8:4, 291–302, https://doi.org/10.1080/17502977.2014.964454

Sadiki, L (2016) 'The Arab Spring: the 'People' in international relations', in: L. Fawcett (ed.) *International Relations of the Middle East* (Oxford University Press), pp. 325–55.

Schneider, V (2012) *'Governance and complexity'*, in: Levi-Faur, D. (ed.) Oxford Handbook of Governance. (Oxford University Press), pp. 129–142.

Smith, RM (2015) *Political peoplehood: The roles of values, interests and identities* (University of Chicago Press).

Starr, F (2013) *Lost Enlightenment: Central Asia's Golden Age from the Arab Conquest to Tamerlane.* (Princeton University Press).

Swyngedouw, E (2004) "Globalisation' or 'Glocalisation'? Networks, territories and rescaling', *Cambridge Review of International Affairs*, 17:1, 25–48, DOI: 10.1080/0955757042000203632

Tucker, B and D Nelson (2017) 'What does economic anthropology have to contribute to studies of risk and resilience?', *Economic Anthropology*, 4:2, 161–172, https://doi.org/10.1002/sea2.12085

Vogelsang, J (2002) 'Futuring: A complex adaptive systems approach to strategic planning', *Practitioner* 34:4, 8–12.

Waever, O (2002) 'Identity, communities and foreign policy', in: Hansen, L and O Waever (eds.) *European integration and national identity: the challenge of the Nordic states*, 20–49 (London, New York: Routledge)

Walker, J and M Cooper (2011) 'Genealogies of resilience: From systems of ecology to the political economy of crisis adaptation', *Security Dialogue*, 42, 143–160, https://doi.org/10.1177/0967010611399616

Wendt, A (1994) 'Collective identity formation and the international state', *The American Political Science Review*, 88:2, 384–396, https://doi.org/10.2307/2944711

White, J (2020) 'The enduring appeal of autocrats', *Cambridge Review of International Affairs*, 33:6, 925–930, DOI: 10.1080/09557571.2020.1838820

Yountae, A (2017) *The decolonial abyss: Mysticism and cosmopolitics from the ruins* (Fordham University Press)

Zebrowski, C (2013) 'The nature of resilience', *Resilience*, 1:3, 159–173, https://doi.org/10.1080/21693293.2013.804672

Decolonising resilience: reading Glissant's *Poetics of Relation* in Central Eurasia

David Chandler

Abstract   *In dominant Eurocentric policy imaginaries, a resilient community is able to self-govern and to autonomously manage risk through becoming more adaptive and responsive to potential threats, mitigating harms and maintaining societal equilibrium; 'bouncing back' rapidly to normal conditions. This paper seeks to move the discussion forward, suggesting alternative framings for the conceptualisation of community practices and understandings as part of the project of decolonising approaches to Central Eurasia. In drawing upon recent works addressing resilience via Édouard Glissant's* Poetics of Relation, *it highlights alternative understandings of resilience which are less subject-centred and more dependent upon* becoming with others in relation. *Crucial to these practices of relationality is the recognition of opacity - the acceptance that uncertainty and unknowability are integral to life processes and provide a vital invitation or opportunity to experiment and adapt through improvisation rather than mechanically responding to feedback effects in ways which close off alternative possibilities for change.*

Introduction

As the editors of this special issue note, our contemporary world of relational entanglement is increasingly captured in discourses of complexity and of the Anthropocene as an epoch in which human activity and nature are mutually imbricated within problems of catastrophic climate change. In this moment, new approaches to governance are evolving with the intention of enabling communities to become resilient. This means, in the parlance of the European Commission, that they are capable of 'bouncing back' and adapting in the face of shocks and disturbances (European Commission 2019; Tocci 2020). The introduction stresses the importance of grasping community resilience as a relational process rather than as some form of fixed goal or fixed set of organisational capacities. The editors suggest that, in the Central Eurasian region, community relations can often draw upon traditions of solidarity and philosophy of good neighbourliness, 'reflected in the enduring notions of *'hamsoya'* (united in shadows); *'wahdat al wujad'* (unity of beings); *'hamdardi'* and *'ham-dili'* (compassionateness, kindness and forgiving)' which go beyond Western (and neoliberal) conceptions of individual autonomy and responsibilisation (Korosteleva and Petrova 2021, Introduction to this special issue).

These practices and understandings provide opportunities to present community resilience in a different register to that found in mainstream EU policy approaches, which tend to be replicated for 'Eastern neighbours' to adhere to. The papers in this collection suggest that connections can be drawn across a range of beliefs and practices which stress community in terms of a *processual becoming-with* rather than in binary Western or Eurocentric framings of the individual rational choice-making subject, which is to be 'nudged' into more communal and sustainable outcomes (Thaler and Sunstein 2009). This paper seeks to underline the importance of charting these distinctions, thus highlighting the importance of non-Eurocentric readings of resilience for developing decolonising approaches to Central Eurasia and challenging the central assumptions of the dominant policy-framings of resilience.

These assumptions are, firstly, that the status quo is the norm, which should be maintained or 'bounced back' to in the face of potential or actual disruption, imagining the future as merely the linear extension of the present. Secondly, there is the assumption that the autonomous choice-making individual is the political model for rationalist decision-making and adaptive behaviour scaled up to the community, regional or state-levels. Thus, community resilience is established upon the basis of an informed, 'empowered' and active citizenry, alert to changing circumstances and able to adapt in order to sustain communities as stable organisational entities. Both of these assumptions concur ontologically with a modernist framing of a world that is made available to be known and ordered by the human as subject: community resilience is thereby an exercise in knowledge and control to reorder or reattain order. It is precisely these aspects that are challenged in this paper, taking the call to decolonise IR approaches to Central Eurasia beyond the empirical recognition that the region does 'not necessarily follow the patterns of development, agency, and state behavior paved by the European experience' (Dadabaev and Heathershaw 2020, 12).

This paper is organised in three main sections. Firstly, the problematic of resilience is set out as presented in the gap between Eurocentric assumptions of community resilience and those located beyond the parameters of the modernist imaginary. Crucial here is the importance of alternative conceptions of relation, such as those forwarded by the Caribbean author Édouard Glissant, who is influential for a number of alternative approaches to resilience in the register of Black, Queer and Decolonial understandings.

In the framing of this special issue, these alternative perspectives of relational processes, of community, and of resilience provide important and relevant insights into how one should approach these questions without imposing the governance mindset of international policy-makers to cases 'that do not necessarily fit within narratives centred on state power and/or socialisation according to Western norms' (Dadabaev and Heathershaw 2020, 3; see also Lottholz et al. 2020). Pursuant to this, the second section follows the lead of Kara Keeling in *Queer Times, Black Futures* (2019), who considers approaches to resilience which hold the future open in contradistinction to the closure of 'bounce back' approaches associated with homeostatic modulation and Deleuzian 'Societies of Control'. This framing helps at getting to the heart of what is at stake in contemporary discussions of resilience as the governance of complexity and contingency. Keeling importantly argues that Glissant's work could be a template for approaches which seek to keep the future open in a non-deterministic way.

In the third section, I engage with An Yountae's *The Decolonial Abyss: Mysticism and Cosmopolitics from the Ruins* (2017) which speaks directly to resilience and community construction from the perspectives of non-Eurocentric understandings, tracing the spiritual and political attempts to grasp uncertainty as positive and enabling from Neoplatonic thought through to the work of Glissant. Yountae challenges the Eurocentric tendency to see the abyss (the unknowability of the outside/other) as facilitating the growth of world history/the subject (as per Hegel and Žižek) and reads Fanon and Glissant as enabling us to 'stay with the trouble' (Haraway 2016) of contingency, opacity and unknowability. Drawing from the positive and enabling readings of Glissant provided by Keeling and Yountae, the conclusion summarizes how it is possible to carve out alternative framings of community resilience which move beyond Eurocentric or 'neoliberal' formulations. Firstly, these alternative approaches seek to hold the future open temporally and spatially, through the understanding of *community resilience as a process of becoming-with*. Secondly, the individual is no longer centred in these processes of *becoming-with*, which inculcate a collective ethos of experimentation, as opposed to a responsibilised subject suborned to a modernist imaginary of empowerment and self-growth.

Resilience

One difficulty that Eurocentric approaches to resilience face is the question of how to address community cohesion and community development in a context where risks and threats are not known in advance. In a world of complexity, a world of relational entanglement, the capacity to be responsive to feedback effects is at a premium (Rist et al. 2014). Resilience approaches that focus on 'bouncing back' tend to assume the goal as stability, taking the world as it is, as a status quo or a given. From this position of fixity in time and space, resilient communities are then imagined to require 'empowerment' or 'capacity-building' so that they are able to sense and respond to changes through quick decision-making, coping through maintaining stability and system functioning. The aim being to minimize the disruption of disasters or conflict and to 'bounce-back' to normal as rapidly as possible. In this framing, the speed of reaction is vital, the faster that problem signs or ('early warning') signals are recognised, evasive or preventive measures can be taken (UNEP 2015). Thus, the more effective resilience systems of detection are, the more perfect the capacities for response and recognition, and the more ingrained or automatic feedback responses can become. External aid is thereby often less about telling communities what to do but how to 'be'; how to organise or institutionalise mechanisms that enable them to see, recognise or register changes which may be indicative of threats or problems (World Bank 2017).

The goal for resilience would be a community which faced little to no disruption, with the development of adaptive measures making the response to feedback increasingly 'real time'. In this framing, transparency and automatic responses are key aspirations as signs and signals of change are rapidly interpreted and reacted to. If stability is the goal, then the automated and stabilising approach of resilience would operate in ontological terms of homeostatic regulation, the regime of cybernetic governance, as a fixed or pre-given goal. But what if resilience, as crisis or threat response, cannot be automated or ingrained in community responsivity? What if problems or disturbances emerge in ways which do not automatically trigger warning signals and alarms? What if returning to 'normal' is not the solution but actually part of the problem? What if community resilience requires a different or more experimental approach? What if resilience requires us to be more open to the world rather than seeking to react and respond in a negative or protective manner?

The approach to community resilience found in the thought and practice of those working outside the assumptions informing institutional and NGO/INGO policy doctrines can provide alternative possibilities. These possibilities are important as they indicate opportunities for rethinking resilience beyond the limitations of the modernist ontology at the heart of Eurocentric approaches (see also Chandler, Grove, and Wakefield 2020). There have been alternative voices, or a minoritarian trend of thinking within the West, which have disputed the assumptions at play in hegemonic policy circles. The need to open up resilience frameworks was perhaps most presciently and cogently argued by the French philosopher Gilbert Simondon, with his conceptualisation of 'metastability' rather than 'stability' – disputing the cybernetic imaginary that societies should be seen as fixed and needing to modulate around the equilibrium (2017). Simondon's reason was that life should be grasped ontologically as fluid and dynamic rather than existing in fixed relations which can be made knowable and transparent. A conception of metastability understands relations as unpredictable, as always in flux, and thereby never fully graspable, always 'coming-into-being' (Simondon 2017, 169). There is thus nothing mystical in understanding life as in relation, as in flux and not fixed, as 'metastable' rather than structured around the goal of stability. As Korosteleva and Petrova highlight in their Introduction to this special issue, the policy goal of equilibrium is problematised when we consider resilience as a process of complex adaptation (see also Orsini et al. 2020). As Simondon notes, the Western ontology of stability and fixed relations reduces life to a mechanistic rather than living existence:

> … stable equilibrium, in which all potential would be actualized, would correspond to the death of any possibility of further transformation; whereas living systems, those which precisely manifest the greatest spontaneity of organisation, are systems of metastable equilibrium; the discovery of a structure is … not the destruction of potentials; the system continues to live and evolve; it is not degraded by the emergence of structure; it remains under tension and capable of modifying itself. (2017, 177)

A similar framing can be seen in C. S. Hollings' highlighting of resilience in ecological systems, in which he outlined his understanding of resilience in relation to the 'adaptive cycle' (1973). Here resilience depends on system openness and adaptability: too much certainty in the reproduction of fixed ways of working and understanding easily leads to system failure. Importantly, for Hollings and his associates, 'ecological resilience' was distinguished from 'engineering resilience', where structures 'bounce back' from stresses, as it presumes the existence of multiple potential stable states or regimes, rather than only one (Gunderson 2000; see also Grove 2018;

Wakefield 2020). A similar call for resilience approaches to move away from the focus on maintaining order and certainty was made by Nassim Taleb in his call for 'antifragility'; in recognition that uncertainty—the complexity of relational entanglements—forces us to be open to the world, whereas modernist frameworks which imagine certainty are necessarily fragile—more vulnerable to unforeseen side effects or changes (2013). We could read a similar understanding in the late Ulrich Beck's assertions of 'world risk society' and the recognition that unintended side effects or 'externalities' often were of more consequence than intended consequences (2009). Uncertainty, assumptions of the inability to know and to automatically respond are crucial to more open framings of resilience which, following Donna Haraway's edict, seek to 'stay with the trouble' (2016).

These are still minority positions and understandings when it comes to the export of policy approaches to Central Eurasia. The reason may be the difficulty of European policy elites in seeing beyond institutional policy needs of uniformity and stabilisation (Bickerton 2015). The basis of contemporary hierarchies, reproduced in policy prescriptions and guidelines, particularly in relation to the EU's eastern 'neighbours', is that the scientific workings and technical expertise of the EU is something to be exported and emulated (Chandler 2010). Understanding life as dynamic and agential, complex and differentiating, calls into question the ontological assumptions underpinning EU managerial expertise. For contemporary Western policy advocates, policy is something to be centralized and regulated from above, something to be benchmarked and box-ticked. It is all too easy to think about resilience as a way of bringing together and merging policy requirements, from defence to social welfare, providing universal frameworks for scaling up 'capacities' and 'empowering' communities. It is also all too easy to consider any other non-equilibrium approaches as non-scientific, speculative or mystical (Thacker 2010), perhaps falling back on traditional Sufi or other monist understandings which allegedly fail to recognise the centrality of the human/nature divide (Shahi 2019).

In an attempt to broaden the discussion in this special issue beyond a potentially essentialising discussion of the merits and drawbacks of the universalist approaches of international policy advocacy and traditional understandings of Central Eurasian community practices and beliefs (see discussion in Lottholz et al. 2020), this paper seeks to draw upon other non-Eurocentric framings and understandings of resilience. It seeks to highlight that the assumption, that relations within communities and of communities to the world should be open rather than fixed, does not imply that there is some sort of life force or non-human transcendental or immanent agency at work. *There is nothing 'backward' or traditional about understanding resilience as a processual becoming in a world of flux and change.* In fact, if communities are to have a future framed in terms of resilience, then it is clear that dominant policy discourses are required to change, regardless of the immediate difficulties policy providers may face. A relational rather than a rationalist ontology offers an alternative conceptualisation, that there can be no goal of self-regulating finality or of resilience as simple adaptation to the status quo. Policy assumptions of transparency, the desire for rapidity and the automation of response, can be construed not as goals but actually as barriers to communities' self-realisation. It is for this reason that non-Eurocentric approaches can be usefully approached for those seeking alternatives to the constrictions of current policy advocacy.

Édouard Glissant intimates what is at stake in a decolonial approach to resilience in *Poetics of Relation* (1997), in which he lays out an alternative approach to highlight the limits of 'reductionism' in much Western thinking of relation. Thus, he argues that Einstein's theory of Relativity does not take a relational ontology far enough and thereby 'is not purely relative' (Glissant 1997, 134). Key is the fact that, for Einstein, '[t]he universe has a 'sense' that is neither chance nor necessity', this provides "guarantees' [both of] the inter-active dynamics of the universe and of our knowledge of it' (Glissant 1997, 134). Thereby: 'Just as Relativity in the end postulated a Harmony to the universe, cultural relativism (Relativity's timid and faltering reflection) viewed and organised the world through a global transparency that was, in the last analysis, reductive.' (Glissant 1997, 135) Thus, for Glissant, there are two 'tendencies' or ontological approaches of understanding relational becoming.

The first approach is the colonial one, appealing to scientific, evolutionary, or underlying cybernetic laws and rationalities of 'interactive life' that 'has become increasingly based on attempts to imagine or to prove a 'creation of the world' (the Big Bang), which has always been the 'basis' of the scientific project' (Glissant 1997, 136–137), enabling a Darwinian evolutionary telos of progress. Despite claims often to the contrary, 'The idea of God is there. And the notion of legitimacy reemerges. A science of conquerors who scorn or fear limits; a science of conquest.' (Glissant 1997, 137) The second approach to relation, on the contrary, tends in:

> ...the other direction, which is not one, distances itself entirely from the thought of conquest; it is an experimental meditation (a follow-through) of the process of relation, at work in reality, among the elements (whether primary or not) that weave its combinations... This 'orientation' then leads to following through whatever is dynamic, the relational, the chaotic—anything fluid and various and moreover uncertain (that is, ungraspable) yet fundamental in every instance and quite likely full of instances of invariance. (Glissant 1997, 137)

Glissant (1997, 142) therefore advocates an alternative approach to knowledge, of *poetics*, challenging universal, generalising or transcendent totalities in its focus on ever 'more stringent demands for specificity.' His approach is a practical one, in which the subject is no longer an observer of relations but always *practically worlding itself* in a concrete embedded and embodied way. This focus upon contextual specificity in practices of 'worlding' or 'becoming' necessarily implies what he calls 'the right to opacity' (1997, 190). The right to opacity would imply that community resilience could be thought via an alternative set of assumptions. For example, that transparency and automated feedback are not as important as the assumptions of indeterminacy, invisibility and lack of knowledge (see also Pugh and Chandler 2021).

The 'right to opacity' is vital to keep communities open to changes which cannot be predicted beforehand and to which there is no necessarily fixed or 'one size fits all' response which can be automated. As Glissant states (1997, 190–191), the notion of 'opacity' highlights 'an irreducible singularity': 'The opaque is not the obscure, though it is possible for it to be so and be accepted as such. It is that which cannot be reduced, which is the most perennial guarantee of participation and confluence.' This approach then may view communities as themselves changing in the ways they see the world and respond to it, allowing for the growth and development of communities as they 'world themselves' in an open set of responsivities rather than closed ones. In such a framing, relations of openness come prior to any closure of a homogenous, fixed or determined identity as the 'norm'. Relations make a resilient community; one based upon the free play of difference, rather than assuming any *a priori* subject. Autonomy is thus a process of becoming-with others but without assuming unity over difference. This is particularly important for the diverse and interlocking communities of Central Eurasia, as Christian Reus-Smit highlights (2018). In fact, it is difference that enables communities to develop and sustain themselves in the face of shocks and setbacks, and which multiply and enable capacities to respond to feedback effects.

As Tiffany Lethabo King notes, in her reading of Glissant, this establishes a 'poetic politics', which can 'conceptualize a kind of 'uncharted' surroundings that are continually made, remade, or unmade' (2019, 8). The key point about Glissant's conception of 'opacity', shared by her use of the 'black shoal', is that this slows and disrupts assumptions of regularity and linearity in dominant Western or Eurocentric approaches, and 'enables other modes of thinking' that 'opens up other kinds of potentialities, materialities and forms' (2019, 8). Rather than the fixed and automated reflexes of Eurocentric forms of resilience thinking or romantic and essentialising imaginaries of traditional communities as similarly fixed with ingrained learnt responses rooted in land and tradition, the notions of opacity and irreducibility enable conceptions of communities of relations which are not bound to the constraints of 'bouncing back' with its assumptions of flat differentiated space and sequential linear time of modernity.

Thus, the contraposition of non-Eurocentric and hegemonic policy approaches demonstrates different understandings or ontologies at play. For dominant Western policy framings, maintaining stability is key, whereas for other community understandings, autonomy and freedom are highlighted through opacity. In the latter case, it is relations not entities which are fundamental and therefore relational openness; communities are thereby in or amongst a world of flux and flows rather than above or separate to a fixed world of things or essences. Community practices generate greater knowledge of relation rather than responding to feedback effects as if in a world of fixed, regular, and repetitive laws of causation. What is required is a culture of openness to the world and not one of closure. Thus, regional forms of resilience can take numerous forms expressing their specific modes of creativity and openness to change, based upon valuing opacity and freedom. Western policies of resilience, understood as automated adaptation to the world, close off and are antithetical to such non-Eurocentric regional understandings.

Poetics and futural openness

In order to think through the logic and implications of non-Eurocentric understandings of community resilience, especially via the methodological framework of poetics (highlighted in studies of regional approaches in this special issue), this section draws upon a close reading of Kara Keeling's 2019 study *Queer Times, Black Futures*, which works with Glissant's key conceptual framings to move beyond the ontological constraints of dominant approaches. Rethinking the shibboleths of linear time and the fetishised understanding of the autonomous subject means that community resilience can be understood and developed in ways which are distinct from the policy doctrines rolled out by the EU and other institutional bodies. Important to note here is an understanding of temporality that does not assume the linear framing of 'bouncing back' to the equilibrium or status quo, nor the subject capable of knowing the world in transparent ways and responding through rational choice to maintain order. Keeling associates an alternative framing with Queer and Black approaches to temporality which are seen to disrupt linear causal understandings and, in so doing, to hold the future open (see also Rao 2020). Chance, disorder and disruption are seen as part of the world, both human and nonhuman, providing capacities for change that should be enabled rather than closed off, avoided or ignored. She argues that 'None of us survive as such; indeed, perhaps, freedom requires we give way to other things.' (2019, ix) This is posited in contradistinction to 'bounce back' understandings that seek to modulate around an equilibrium, using the sorts of algorithmic technology presciently engaged in Deleuze's 'Postscript on Societies of Control' (2019, 12; see also Deleuze 1995)

Rather than dominant resilience approaches of automated rationalist techniques of reaction to changes and disturbances in terms of returning to the status quo, 'holding uncertainty open, critical theory, poetry, dance, literature, philosophy, music, and other creative sonic phenomena can continue to feed thought' and communal imaginaries (2019, 15). For Keeling, 'Queer temporality' refuses the linear form of modulation around the norm and is 'violent, material, and excessive to the management and control of sociability … 'queer' remains an active and energetic reservoir for connection, affiliation, and experimentation' (2019, 18–19). In fine, an alternative reading of resilience emerges through inversing the relation between world and community resilience proffered in Eurocentric framings that prioritise stability. Resilience understood as futural openness prioritises disruption over order in the sense of understanding that the uncertain, the uncontrollable and the unknown can be liberatory rather than oppressive or problematic. This is why, for Keeling, Glissant's conception of 'opacity' is central for a 'politicized cultural strategy' invested in 'Black Futures' and 'queer temporality, which resists the Eurocentric or Western 'requirement for transparency' (2019, 31):

> Glissant argues it is important for marginalized groups to 'insist' upon remaining opaque to the terms, languages, and logics of dominant groups. Insisting upon opacity acknowledges the co-existence of systems of signification and valuation alongside, yet inaccessible to dominant ones. Within this context, 'unaccountability' marks a refusal to be bound to dominant standards of measure, recognition, and evaluation. (2019, 46)

Opacity, the capacity to hold the world without transparency, without assimilating newness and difference to what exists and is known, without reducing signs and signals of change to pre-set patterns and meanings nor individual entities to categories of comparison, commensuration and equivalence, enables another world to come into being (Keeling 2019). More than just a practice of open relationality to the world, 'the right to opacity' also challenges the requirements of external policy advocacy, seeking to monitor and benchmark community 'development' or 'capacity-building' (see, for example, Korosteleva and Flockhart 2020). This understanding–that the strict separation between human and nature or knowing subject and transparent object, limits alternative possibilities of becoming resilient–offers opportunities for communities to resist and to refuse the universalising metrics of EU guidance and regulatory control.

This inversion of problem and norm via the understanding of the importance of disruption in thinking creatively/poetically, is derived from Black and Queer experiences of the constraints of the norm, thereby opening up a problematic through which these ideas and practices can be seen as having a broader impact on how resilience is understood today. One notable example is the governmental responses to the COVID-19 crisis throughout 2020, in which popular pressure has resisted attempts to 'bounce back' and 'return to normal' and has highlighted that the disruption has created a wide range of alternative possibilities. These range from how we think human relations to animals (Kothari et al. 2020), the problems of dependence on cheap labour and zero contracts, the prevalence of race and gender inequality of outcomes, the health implications of housing inequalities and the underfunding of the health service (Horton 2020), to the need to rethink environmental impacts of current working practices. 'Normal' will, and should, never be seen the same way as it was before the crisis (Chandler, Grove, and Wakefield 2020). This framework of thinking fits well with the late Ulrich Beck's understanding of 'emancipatory catastrophism', where crises enable new relational entanglements to become visible and spur governance interventions in response to new understandings and awareness of how issues are interconnected (Beck 2015).

This opening up of alternative future possibilities, diverging from linear expectations and predictions based upon the past, could be construed as a 'queer temporality', one that brings into question the Eurocentric assumption of a linear time of 'progress', where instability is merely a temporary pause on the journey of 'lessons learned' as life continues upon its predetermined path. Rather than one line of linearity, breaks and disruptions open up the possibilities of 'alternative worlds' that cannot be imagined and their courses plotted beforehand. Thinking along these lines begins to open up opportunities that can be understood to exist in the present but that are unseen or require disruptions to become actualised. For Glissant, this understanding of the present as always open rather than a closed or predetermined moment in a linear chain, was largely shaped by the Caribbean experience of the displacement of transatlantic slavery and colonial domination (Drabinski 2019). This creates a relationship to time and space which breaks with the fixed understandings of community, often imagined in Western policy doctrines of resilience, disrupting the fixed positionality of the subject at the centre of the world. This experience of dislocation is also one of profound interconnection, as Keeling writes:

> Homeless is our home. We carry the abyss that Édouard Glissant characterized so well. For Glissant, the Middle Passage of the transatlantic slave trade and the formation of 'the new world' mark an apocalyptic catastrophe. We are forged in its wake. With specific

reference to those who can be identified as Caribbean, Glissant explains: 'The abyss is also a projection of and a perspective into the unknown... This is why we stay with poetry... We know ourselves as part and as crowd, in an unknown that does not terrify. We cry our cry of poetry. Our boats are open, and we sail them for everyone.' (2019, 54)

Open in and to the world, alternative framings of resilience thereby understand themselves as always in *the middle of processes of inter-relation* rather than standing opposed to or external to them. Similar understandings can also be seen as central to some of the community belief systems in the Central Eurasian region (see for example, Green 2012; also Nurulla-Khodzhaeva in this special issue). Governing and responding in a manner of being open to and within the world of necessity becomes more experimental and spontaneous or improvisational. Without the props of certainty and of linear causality, where there are assumptions that the same actions produce the same outcomes, independently of time and space, it is necessary that responses are iterative or recursive (2019, 55). It is this break that Beck described as 'metamorphosis', a 'change in the conditions and understandings of change' (2015, 76) which 'challenges the way of being in the world, thinking about the world and imagining and doing politics (2015, 78). In this framing, even disasters can be 'emancipatory', bringing new relations into being. Beck uses the example of Hurricane Katrina that devastated New Orleans and the Louisiana coast in 2005:

> Until Hurricane Katrina, flooding had not been positioned as an issue of environmental justice—despite the existence of a substantial body of research documenting inequalities and vulnerability to flooding. It took the reflection both in publics and in academia on the devastating but highly uneven 'racial floods' of Hurricane Katrina to bring back the strong 'Anthropocene' of slavery, institutionalized racism, and connect it to vulnerability and floods. (2015, 80)

Understanding community resilience in a world of relational entanglement, for Beck, could be emancipatory as 'a new normative horizon' emerges (2015, 80), a political horizon that is set by the world of unknown effects and side effects of relational interaction. However, it is important to understand that it is not the disaster itself that makes a difference but the attitude towards it, the pre-existing dispositions that can enable disruptions to have positive impact by making the most of the break in linear temporal assumptions that 'normal' should be 'bounced back' to as a goal.

Black, Queer and Decolonial sensibilities are often central to thinking and developing alternative approaches to resilience as there is less at stake in assumptions that modulating around the 'norm' is desirable. The connection between disruption and the problem of institutionalised racism was, in fact, already made nearly 100 years prior to Beck, by the American sociologist, author and activist W. E. B. Du Bois. As Keeling points out, in his short story 'The Comet', Du Bois argued that a natural disaster would be necessary to shake America out of its racial 'normality'. As Keeling states: 'In Du Bois's story, a natural disaster precipitates a temporary suspension of the terms through which present reality congeals, thereby creating the conditions under which a Black man and a white woman might acknowledge a shared humanity.' (2019, 62) What this approach to the abnormal, the unthought or the (previously) impossible tells us, by inversing the 'normal' and the abnormal or disruptive, is that community resilience built upon assumptions of fixity and linear continuation can be counterproductive: problematically closed in terms of existing injustices and inequalities as much as to the effects of larger social and environmental interactions.

The world becomes stranger and more uncertain, but this opacity acts as an invitation to experimentation and improvisation without linear assumptions. Giving way to things, being responsive to the world, is not a matter of reaction and defensiveness, of protecting the status quo, but of being open to the other, to alterity and to the unknown. How communities can do this in order to be sensitive to feedback effects, i.e. to the changing world around them, is thus not through mechanised responses but rather through poetics, through feelings and emotion, through ways of honing 'capacities to be affected' (2019, xii; see also Latour 2004). Paying attention to differences and changes means becoming focused more upon the unknown rather than the known, can thereby enable creativity beyond community 'common sense', structured around past habits and modes of being. Keeling, draws upon Afrofuturist jazz musician Sun Ra to argue that uncertain or unknown disruptive forces can be grasped as 'poetry from the future' or invitations to think the world differently:

> Sun Ra points toward the ways that whatever escapes or resists recognition, whatever escapes meaning and valuation within our community crafted structures of valuation and signification, exists as an impossible possibility within our shared reality (however that reality is described theoretically) and therefore threatens to unsettle, if not destroy, the common sense on which that reality relies for its coherence. (2019, 62)

Poetics is thereby an approach to alterity that seeks to use this in a way that can disrupt habitual responses and perceptions. Keeling argues that in contemporary thinking about the crisis of modernist structures of thought, with its binary and reductionist separations between human and nature and thought and being, oriented around the imaginary of linear time and the autonomous, self-determining rational subject, contemporary Western thinking is increasingly drawn to the radical tradition of thinking from the outside of Black thought. She argues that the work of Simondon and other continental thinkers, in challenging Eurocentric subject-centred approaches, in their emphasis on relation rather than the fixity of essences and entities, very much chimes with concepts developed in Black diasporic thought, traditions and practices (2019, 70). For Glissant, Caribbean or Black diasporic thinking and practices lacked the fixed rootedness of Western or Eurocentric conceptions of time and space as the trauma, violence and uncertainty of the Middle Passage, chattel slavery and coloniality removed pre-existing grounds of community and imposed the necessity of being and becoming with others through improvisation and new shared ways of being.

Thus, community lives and livelihoods can be grasped as lived 'in the break' from groundings of certainty (Moten 2003) and the structured binaries of a modernist ontology. This tradition of radical and experimental thought, shaped by violence and uncertainty and the lack of possibility of secure subjecthood, Glissant understood as one of 'abyss', of working from and within radical rupture (2019, 78–79). In terms of approaches to the Central

Asian region, it is important to note that the understanding of 'Blackness' as an approach, which differs radically from the Eurocentric subject-centred tradition, is not one fixed upon epidermalized divisions: '"Blackness" itself becomes mobile once it is understood in terms of its characteristic cultural form of repetition, rather than as a set of essential qualities of Black people.' (2019, 160–161) Blackness is articulated as a 'kind of 'sensibility', rather than as a property of any particular body or group' (2019, 161). 'Abyssal' sensibilities, key to alternative approaches to resilience, are not grounded in any linear understandings of the future but as Keeling argues:

> As I have been suggesting throughout this book, rather than conceptualizing 'the Black's' lack of perceptible future as a problem to be solved or a crisis to be addressed, or a cause for pessimism or optimism, it might be understood as one of the crucial operations of what we might here grasp as the cut of Black existence: it might cleave an opening in the present order of meaning and being through which another structure, another world, perhaps might be 'preciously assembled'. (2019, 174)

Therefore, a poetics of relation is understood to open up alternative possibilities for community resilience based upon understanding relation as something not fixed and potentially automatic, but as something fluid and never fully graspable (2019, 196). Keeling argues that such poetics is necessarily speculative, for it not only 'acknowledges the dense entanglement of matter(s)', but also 'thrives on surprises and accidents' (2019, 199). We can see here, then, a clear alternative formulation of community resilience as a story of affirmative change and adaptation, but one that does not centre itself around fixed understandings of time and space. Discussion within Keeling's work of 'futures' should therefore not be conflated with a modernist or linear framing and could more precisely be grasped as 'futural' imaginaries which hold potentiality open, extending the possibilities of abyssal or speculative thought and practice, open to the world.

Decolonial relationality

Reading An Yountae's *The Decolonial Abyss: Mysticism and Cosmopolitanism from the Ruins* in conjunction with Keeling's work enables us to consider alternative conceptualisations of community resilience potentially grounded in a struggle for existence, which is abyssal or non-ontological, lacking the modernist grounding assumptions of the a priori subject and world as knowable object. Yountae draws from a Black diasporic literature, particularly Martinican thinkers Frantz Fanon and Édouard Glissant, to articulate what he sees as 'the double work of the abyss that dissolves the self and opens up possibility' (2017, 5). He is very clear that the purpose is to question the understanding of resilience as 'bouncing back' to a pre-existing line of progress in a linear narrative centred around a modernist ontology of the pre-given subject, be this the individual, the community, the region or the state. The central question posed is how the disruption or crisis, considered above, may serve as an opening to alternative conceptions of self and world that can be meaningfully conceptualised without restoring a modernist ontology of subject and linear time:

What kind of future does this crack open? Or does it open a future at all? For the universalizing accounts of dialectical becoming might certainly open a future, but a future perhaps all too familiar to us: one that does not break from the genealogy of the old Christian cosmopolitan world order that keeps reproducing itself each time with a different name: modernity, capitalism, liberal democracy, postmodernity, globalisation, and so forth. (2017, 5)

Yountae argues that the telos of progress, the process of learning from errors that marks the transcendental subject of European philosophy, does not hold the future open but is a 'totalitarian' closing of the imagination (2017, 5; see also Moten 2018). He clarifies that the abyss, understood as a world lacking fixity and causal determinacy should not be conflated with the void, lacking values and capacities for affect (2017, 9). The indeterminacy of the abyss has enabled a long philosophical tradition from Neoplatonism to today, which has remained sceptical of foundationalist thought and operated from the margins of modernist frameworks, often seen as irrational or mystical. However, in our contemporary context, abyssal thought has been mainly associated with the resilience and resistance of colonised subjects in a struggle to articulate new frameworks of meaning and possibility, beyond the 'normal' world of trauma, loss, hierarchy and exclusion.

Thus, as noted above, two ontological framings of resilience emerge. These could be read in terms of a 'colonial' or hegemonic form, which could be grasped in terms of 'the Hegelian journey of dialectical becoming, character-ized by the enigmatic resilience of the [rational] subject who reconstructs itself despite constant failures' and a 'decolonial form' epitomised by the work of Édouard Glissant (2017, 23). For Glissant, poetics rather than rationalism works to reconstruct the self and to overcome the traumatic loss of epistemological certainty; the abyss itself is the 'groundless ground' that enables the self's relational becoming (2017, 24). Poetics as a practice enables a subject to be conditioned by alterity; this is precisely because the self-understanding of the subject is that it lacks self-sufficient grounds, any form of essence contained with itself, and is therefore necessarily conditioned by the other. If we think back to the reference to Ulrich Beck in the previous section, this can be seen in his understanding of the other as setting the emancipatory 'new normative horizon' (2015, 80), rather than having pre-set goals or behaving in an arbitrary way.

It can easily be argued that the experience of existential loss, of a dissolution of grounding frameworks of meaning and traditional certainties share much with our contemporary moment of the Anthropocene and climate catastrophe. Contemporary discussions of resilience are, in fact, framed by the loss of traditional policy certainties and an awareness that contemporary entanglements put to question hierarchical assumptions of power and agency (Chandler 2014). In this context, the concept of the abyss—an awareness that the world is one of contingency and indeterminacy—enables a reconceptualisation of the modern subject. In fact, alternative approaches to resilience argue that this reconceptualisation of the subject is necessary for the development of community resilience. As Yountae states:

The self who is undone in the encounter with the abyss, that is the pre-abyssal self, lives with a misguided consciousness. Without having faced or embraced the vertiginous depths beneath the precarious ground of its being, this self views itself as coherent and independent. I am here referring to the self who operates in clearly demarcated binaries and boundaries... Conversely the new self... understands its nature not as an immutable substance but as multiple, fragmented, and always-in-becoming... a creolized self that finds its truth in the never-ending, pluri-singular acts of becoming in relation to the other; [moving] from the self living with a teleological cosmology to a self who understands the end as a new beginning. (2017, 14–15)

The 'creolized' Self is distinct as an individual, a unique and particular social and historical product, but it also lacks any conception of coherent or defining essence, which can somehow become a transcendental marker, placing them in a hierarchical or subordinate relation to others. This 'fragmented', fluid conception of the subject can be seen to have developed in response to the exclusion of racial and colonial constructions of the subject, which universalise (overrepresent) the Eurocentric conception of 'Man' as rational and autonomous (Wynter 2003). As Fanon argued, for the colonial subject there could be no conception of transcendence or linear growth through the abyss of colonial enslavement and subordination (2017, 77, 99; see also Fanon 1986, 112, 219). The abyss of indeterminacy and loss is not a mystical experience, but an ongoing social and political reality for Black diasporic thought. The traumatic experience of loss of chattel slavery and the Middle Passage is the 'paradoxical temporality' drawn out by Glissant, in that the dislocation, deportation and mass deaths of transatlantic slavery removed people from their past attachments and identities, loss was metaphysical as well as physical (2017, 88).

In the destructive reduction of people to 'flesh' and their 'reinscription' as chattel properties without human 'selves' (Spillers 2017), in the hold of the slave ship, the slave auction, and the plantation, out of these 'groundless grounds'—the 'demonic grounds' of Katherine McKittrick (2006)—an alternative or abyssal subject can be understood to have emerged. Yountae argues, 'Glissant finds in the gaping depth of the colonial abyss a womb that gives birth to a new world, a new people whose mode of being find expression in relation and becoming rather than the static terms of essence and being' (2017, 89). The colonial abyss is then the 'groundless ground' on which those denied selfhood and later full admittance to social equality struggled to find coherence and meaning in ways other than those of the 'normal' denied them by racialised exclusion.

Yountae, in his stress upon the historical weight of living after 'the end of the world', living in the concrete circumstances of loss of certainty and foundational groundings, makes clear the difference between the abyssal thinking of Glissant and apparently similar attempts to go beyond a modernist ontology in Continental philosophy, such as those of Gilles Deleuze. The historical weight of traumatic loss and the oppressive denial of rights and privileges (seen as normal) do not enable an experience of abyssal contingency and uncertainty as liberating the self: there can be no joyful 'lines of flight' for nomadic subjects able to pick and choose identities and to travel and 'transgress' borders and boundaries (2017, 105; see also Leong 2016; Jackson 2015). For Yountae:

The undeveloped trop of relation in Deleuze (Braidotti) and Hegel (Zizek) becomes, in Glissant, the very material with which he transposes the void of loss, the painful middle of fragmented history. We could say that one important difference between the boundless freedom of nomadic ontology and creolized freedom lies in relation. The creolized self and her freedom are conditioned—and enabled—by relation. A limitless horizon of being opens up in the inexhaustible mystery of the other—and in the illimitable webs of solidarity with unknown others. (2017, 117)

The horizon is 'limitless' because the practice of relation is other-directed, rather than oriented around a modernist ontology of the interest-bearing individual subject seen as existing prior to and independently of relation. The important point to note, in terms of this special issue and to the broader project of decolonising approaches to the Central Eurasian region, is that relation is a product of practice rooted in social and historical circumstances. Queer, Black and Decolonial futural practices, or any other forms of non-modernist speculative thought, cannot be chosen or picked up as an alternative set of governmental practices to be turned into policy briefs for community capacity-building and resilience. At the same time, it is possible that much can be learned from alternative understandings, alternative cosmologies and alternative ontologies of community resilience, especially in our contemporary moment. This is pressingly apparent when firstly, traditional understandings of the resilient subject as autonomous and self-reliant are increasingly questioned and, secondly, when 'bouncing back' to 'normal' or continuing existing trajectories of 'progress' are seen to be problematic and contributory factors in the problems that communities are facing.

Conclusion

For the architects of adaptive governance, community resilience is essential to finding new ways of coping in a world that is threatened by climate tipping points. This paper has highlighted the limits of Eurocentric understandings of resilience where communities, imagined to be 'at risk' or 'vulnerable', are encouraged to practice reactive adaptation in order to 'bounce back' to their normal development trajectories. Resilience is thereby understood as the capacity to be aware of and responsive to feedbacks, acting rapidly in recognition of potential threats and opportunities. It would hardly be novel to flag up how this 'neoliberal' framing threatens to naturalise or romanticise community strategies for coping at the edge of crisis, promoting self-responsibility as 'self-determination' and 'empowerment' (Chandler and Reid 2016). This paper has attempted to move beyond the critique of resilience policy prescriptions to engage with a growing attention to alternative possibilities for rethinking community resilience. These alternative possibilities are suggested through the concrete historical experience of Black diasporic thought and practice and are indicative of a much broader range of non-Western understandings and experiences beyond the limits of Eurocentric or Western approaches.

Central to the argument forwarded here has been the analysis of non-Eurocentric approaches which challenge and seek to move beyond two assumptions central to dominant policy-framings of resilience. Firstly, the assumption of linear temporality, with its telos of 'development' or 'progress' as a fixed trajectory which needs to be restored. Whereas neoliberal approaches seek to 'bounce back' to preserve the status quo, alternative approaches take more seriously the contingency and unknowability of our contemporary condition. These alternatives suggest that automating feedback responses cannot work when addressing novel threats and conditions, while also expressing a desire to open rather than close possibilities provided by disruptions, enabling the 'normal' conditions to be rethought, rather than reinforced, and thereby opening up alternative futural imaginaries. Secondly, assumptions of the autonomous and self-governing subject–whether at the level of the individual or the community–have been challenged on the basis that a world of contingency and uncertainty implies the need for a radical openness to the world, decentring the modernist subject and understanding the self as continually in the process of construction through communities of relation, becoming-with others.

Disclosure statement

No potential conflict of interest was reported by the author.

References

Beck, U. 2009. *World at Risk*. Cambridge: Polity.

Beck, U. 2014. "Emancipatory Catastrophism: What Does It Mean to Climate Change and Risk Society?" *Current Sociology* 63 (1): 75–88.

Bickerton, C. 2015. *European Union Foreign Policy: From Effectiveness to Functionality*. Basingstoke: Palgrave.

Chandler, D. 2010. "The EU and Southeastern Europe: The Rise of Post-Liberal Governance." *Third World Quarterly* 31 (1): 69–85.

Chandler, D. 2014. *Resilience: The Governance of Complexity*. Abingdon: Routledge.

Chandler, D. 2020. "Anthropocene Authoritarianism (Critique in Times of Corona)." *Critical Legal Thinking*, 9 April.

Chandler, D., Grove, K. and Wakefield, S., eds. 2020. *Resilience in the Anthropocene: Governance after the End of the World*. Abingdon: Routledge.

Chandler, D. and Reid, J. 2016. *The Neoliberal Subject: Resilience, Adaptation and Vulnerability*. London: Rowman & Littlefield.

Dadabaev, T. and Heathershaw, J. 2020. "Central Asia: A Decolonial Perspective on Peaceful Change." In *The Oxford Handbook of Peaceful Change in International Relations*, edited by T. V. Paul, D. W. Larson, H. A. Trinkunas, A. Wivel, and R. Emmers, 1–17. Oxford: Oxford University Press.

Deleuze, G. 1995. "Postscript on Control Societies." In *Negotiations: 1972–1990*, edited by G. Deleuze, 177–182. New York: Columbia University Press.

Drabinski, J. E. 2019. *Glissant and the Middle Passage: Philosophy, Beginning, Abyss*. Minneapolis: University of Minnesota Press.

European Commission. 2019. "Fact Sheet: Resilience." February. https://ec.europa.eu/echo/what/humanitarian-aid/resilience_en

Fanon, F. 1986. *Black Skin, White Masks*. London: Pluto Press.

Glissant, E. 1997. *Poetics of Relation*. Ann Arbor: University of Michigan Press.

Green, N. 2012. *Sufism: A Global History*. Chichester: Wiley-Blackwell.

Grove, K. 2018 *Resilience*. Abingdon: Routledge.

Gunderson, L. H. 2000. "Ecological Resilience—In Theory and Application." *Annual Review of Ecology and Systematics* 31: 425–439.

Haraway, D. 2016. *Staying with the Trouble: Making Kin in the Chthulucene*. Durham: Duke University Press.

Holling, C. S. 1973. "Resilience and Stability of Ecological Systems." *Annual Review of Ecology and Systematics* 4: 1–23.

Horton, R. 2020. "Offline: COVID-19 Is not a Pandemic." *The Lancet*, 396, 874.

Jackson, Z. I. 2015. 'Outer Worlds: The Persistence of Race in Movement "Beyond the Human."' *GLQ: A Journal of Lesbian and Gay Studies* 21 (2/3): 215–218.

Keeling, K. 2019. *Queer Times, Black Futures*. New York: New York University Press.

King, T. L. 2019. *The Black Shoals: Offshore Formations of Black and Native Studies*. Durham: Duke University Press.

Korosteleva, E. A. and Flockhart, T. 2020. "Resilience in EU and International Institutions: Redefining Local Ownership in a New Global Governance Agenda.' *Contemporary Security Policy* 41 (2): 153–175.

Korosteleva, E. A. and Petrova, I. Forthcoming. "From 'the local' to 'the global': What makes communities resilient in times of complexity and change?" *Cambridge Review of International Affairs*.

Kothari, A. et al. (2020) "Coronavirus and the Crisis of the Anthropocene.' *The Ecologist*, 27 March.

Latour, B. 2004. "How to Talk About the Body? The Normative Dimension of Science Studies." *Body & Society* 10 (2–3): 205–229.

Leong, D. 2016. "The Mattering of Black lives: Octavia Butler's Hyperempathy and the Promise of the New Materialisms." *Catalyst: Feminism, Theory, Technoscience* 2 (2): 1–35.

Lottholz, P., J. Heathershaw, A. Ismailbekovac, J. Moldalieva, E. McGlinchey, and C. Owen. 2020. 'Governance and Order-Making in Central Asia: From Illiberalism to Post-Liberalism?" *Central Asian Survey* 39 (3): 420–437.

McKittrick, K. 2006. *Demonic Grounds: Black Women and Cartographies of Struggle*. Minneapolis: University of Minnesota Press.

Moten, F. 2003. *In the Break: The Aesthetics of the Black Radical Tradition*. Minneapolis: University of Minnesota Press.

Moten, F. 2018. *The Universal Machine (Consent not to be a Single Being)*. Durham: Duke University Press.

Orsini, A., P. Le Prestre, P. M. Hass, M. Brosig, P. Pattberg, O. Widerber, L. Gomez-Mera, J.-F. Morin, N. E. Harrison, R. Geyer, and D. Chandler. 2020. "Complex Systems and International Governance." *International Studies Review* 22 (4): 1008–1038.

Nurulla-Khodzhaeva, N. forthcoming. "Imitated and Real Resilience of Sufi-hamsoya." *Cambridge Review of International Affairs*.

Pugh, J. and D. Chandler. 2021. *Anthropocene Islands: Entangled Worlds*. London: University of Westminster Press.

Rao, R. 2020. *Out of Time: The Queer Politics of Postcoloniality*. Oxford: Oxford University Press.

Reus-Smit, C. 2018. *On Cultural Diversity: International Theory in a World of Difference*. Cambridge: Cambridge University Press.

Rist, L., A. Felton, M. Nystrom, M. Troell, R. A. Sponseller, J. Bengtsson, H. Osterblom, R. Lindborg, P. Tidåker, D. G. Angeler, R. Milestad, and J. Moen. 2014. "Applying Resilience Thinking to Production Ecosystems." *Ecosphere* 5 (6): art73-11

Shahi, D. 2019. "Introducing Sufism to International Relations Theory: A Preliminary Inquiry into Epistemological, Ontological, and Methodological Pathways." *European Journal of International Relations* 25 (1): 250–275.

Simondon, G. 2017. *On the Mode of Existence of Technical Objects*. Minneapolis: University of Minnesota Press.

Spillers, H. J. 2017. "Mama's Baby, Papa's Maybe: An American Grammar Book." In *Afro-Pessimism: An Introduction*, edited by F. B. Wilderson. 91–122. Minneapolis: Racked & Dispatched.

Taleb, N. N. 2013. *Antifragile: Things that Gain from Disorder*. London: Penguin.

Thacker, E. 2010. *After Life*. Chicago: University of Chicago Press.

Thaler, R. H. and Sunstein, C. R. 2009. *Nudge: Improving Decisions about Health, Wealth and Happiness*. London: Penguin.

Tocci, N. 2020. "Resilience and the Role of the European Union in the World." *Contemporary Security Policy* 41 (2): 176–194.

United Nations Environment Programme (UNEP). 2015. *Early Warning as a Human Right: Building Resilience to Climate Related Hazards*. Nairobi, Kenya: United Nations Environment Programme.

Wakefield, S. 2020. *Anthropocene Back Loop: Experimentation in Unsafe Operating Space*. London: Open Humanities Press.

World Bank. 2017. *Unbreakable: Building the Resilience of the Poor in the Face of Natural Disasters. Climate Change and Development*. Washington, DC: World Bank.

Wynter, S. 2003. "Unsettling the Coloniality of Being/Power/Truth/Freedom: Towards the Human, After Man, Its Overrepresentation—An Argument." *CR: The New Centennial Review* 3 (3): 257–337.

Yountae, A. 2017. *The Decolonial Abyss: Mysticism and Cosmopolitics from the Ruins*. New York, NY: Fordham University Press.

From 'Westlessness' to renewal of the liberal international order: whose vision for the 'good life' will matter?

Trine Flockhart (iD)

Abstract *The concept 'Westlessness' suggests that the liberal international order (LIO) is in crisis because there is today 'less West' in the world, and the West is itself becoming less Western. The article asks if, and if so how, the LIO can remain resilient and salient in the face of growing 'Westlessness' and contestations against the liberal order's values and institutions. It suggests that the big challenges will be for the LIO to adapt to give the order a broader appeal than is currently the case, whilst doing so without losing what is understood to be the essence of the LIO. The article suggests that this may be possible through a focus on visions for the good life as 'trading zones' in which dissimilar groups can find common ground whilst simultaneously disagreeing on their general outlook. However, the approach necessitates acceptance that perhaps the LIO has for too long been characterised by 'Westfulness' and that 'less West' in the world could be a step towards a more inclusive international order with space for different visions for the good life.*

When the foreign policy and security establishment of the liberal international order (LIO) met at the prestigious Munich Security Conference in February 2020, the theme of the conference was the concept 'Westlessness'. The introduction of the concept reflected a widespread concern within the establishment of the American-led LIO that not only was there 'less West' in the world today, but also that the West was itself becoming 'less Western' and more contested both from within and from without (Bunde et al. 2020). The concept was intended to generate debate about what it means to be 'Western' and was widely expected to become a catchword for 2020 by summarising growing concerns about the crisis of the LIO and about how an emerging, more diverse global international society might be ordered in the future. Although the call for debates about 'Westlessness' was usurped by more immediate concerns in the shadow of the COVID-19 pandemic, the issues raised in Munich will also be important after the pandemic. These issues question if, and if so how, the resilience of the LIO can be maintained notwithstanding the retreat of the West, how contestations against the liberal order's values and institutions can be met, and how its appeal to communities that do not necessarily share the Western and liberal value base of the order may be enhanced in preparation of a less 'Western' global international society. In this process, one of the big challenges will be for the LIO to adapt and to renew its value base to not only give the order a broader appeal than is currently the case, but also importantly to do so without losing what is understood to be the essence of the order. Doing so will involve a delicate balancing act with the potential for a fundamentally renewed—more inclusive and more culturally rich—international order but also with considerable risks to the future integrity of the LIO.

In this article, I pick up on some of the issues raised by the term 'Westlessness'. However, rather than treating contestations and 'Westlessness' as purely a problem, I start from the position that perhaps the problem has for too long been 'Westfulness' and a lack of engagement with the contestations against overtly West-centric power, principles and practice. From this perspective, a situation of 'less West' in the world could be read as a welcome and overdue first step towards inclusivity and openness that leads to a more culturally diverse international order by also including those who thus far have occupied a marginal position vis-à-vis the liberal international order. The communities in Central Eurasia that are the subject of this special issue are in such a position. The development of their relationship with the LIO and its institutions has been hampered by their position in the geographic, ethnic and cultural spaces of the borderlands between Europe and Asia, between Christianity and Islam, between tradition and modernity and between democracy and autocracy. This position has made it difficult for them to find a comfortable fit between their local needs for sustaining their cultural distinctiveness and their pragmatic political ambition for progress and prosperity through a closer affiliation with the order—a situation that by no means is restricted to the communities of Central Eurasia. The article argues that although the current crisis in the LIO is deeply unsettling for the West, crisis and contestation can be a powerful motivation for adaptation and renewal, which can in effect facilitate more long-term resilience. Decision-makers could do well to remember that contestation is more than a latent condition of discontent to be avoided. It is an important social practice that discursively expresses disapproval and entails objection to specific issues that matter to people, and which need to be addressed (Wiener 2017). Practices of contestation are therefore a key component of the reflexive and transformative dynamics that convey inspiration for contemplating change and adaptation (Dunne and Reus-Smit 2017, 34).

The current challenge is that 'Westlessness' implies contestation against the foundational values of the liberal order, which after all are what hold any community—including international orders—together. Addressing issues of contestation and restoring the order's resilience without undermining the deeper social fabric of the order is therefore not an easy task. The danger is that the necessary adaptation for addressing contestations and maintaining resilience might either increase the LIO's inclusivity but in the process weaken the order's cohesion, or alternatively the LIO might maintain its Western and liberal value base but fail to address the contestations and its fading salience in a growing number of constituencies. The article seeks to uncover the conditions for succeeding in the fine balancing act that is required between adaptation and renewal of the values, norms, rules and practices of the LIO to make it more resilient, whilst maintaining a clear alignment with those parts of the deeper social fabric of the LIO that make it what it is—namely, its vision for the future and its shared sense of what constitutes the 'good life'. In this process, the questions are unavoidably going to not only be what the essence of the LIO is, but also whose vision for the future and whose conception of the good life will matter in its adaptation and renewal. This article is, on the one hand, an expression of deep concern about the future of the LIO and, on the other hand, a more optimistic exploration of the possibilities—rather than threats—brought about by the condition of 'Westlessness'.

The article proceeds in four main steps. In the first step, the article focuses on resilience as a practice of self-governance in response to change and contestation and how practice, norms and rules can be adapted, whilst the deeper social fabric of a society is thought to be much less malleable. In the second step, I draw attention to the concept 'the good life' and how it has been expressed in the LIO. I emphasise that the inclusion of the good life should be an additional layer for expressing the purpose of a social entity or community, with reference to the more emotional and softer aspects of life that are often not included in political analysis. In the third section, the article zooms in on the building blocks of all social communities—here conceptualised as the constitutive parts of ideal-type ordering domains. Doing so enables a less 'cluttered' view of the complex workings of all social domains to better understand in which parts of the LIO adaptation and renewal may be needed and possible. The article demonstrates how different conceptions of the good life can play an important role both as a 'trading zone' across dividing lines and as a non-confrontational way of expressing social distinctiveness. In the fourth section, and by way of conclusion, the article turns to consider the prospects for the LIO's *renewal* as a more inclusive and culturally diverse order in which different conceptions of the good life can co-exist as 'trading zones' for finding common ground to meet the many challenges. This is arguably the primary purpose of an international order. The article concludes that contrary to the worries expressed in the discussion about 'Westlessness', cultural and value-based diversity is the lifeblood of orders rather than their weakness.

Resilience, adaptability and meta-stability

The overall focus of this special issue is 'resilient communities'. Resilience is understood both as a specific quality of a system or entity, and perhaps more controversially, as a practice of self-governance to remain 'fit for purpose' in the context of pervasive change, the ubiquity of crises, and the inescapable certainty that life is uncertain (Korosteleva and Flockhart 2020; Korosteleva and Petrova 2021). The understanding of resilience that is forwarded in this special issue is different from the neoliberal discourse on resilience that is common within the International Relations discipline (Chandler 2014; Korosteleva 2020). The understanding of resilience in this issue also differs from the traditional understanding of resilience originating in the fields of ecology and engineering that treats resilience as the systemic and material quality of a system understood as the system's or entity's 'ability to absorb change and disturbance and still maintain the same relationships' (Holling 1973, 14). Although it is important to know that a system such as an oceanic habitat or a loadbearing steel lintel can 'bounce back' to its original form after a crisis, such 'bouncing back' may not safeguard the longer-term ability to remain 'fit for purpose' in a social and human context. In the social world, whilst some situations may indeed require the ability to 'bounce back', others will require the ability to 'jump ahead' to adapt or transform. Therefore, although the traditional perspective is intuitively common-sensical and reified in language that equates 'resilience' with 'robustness' or 'resoluteness' in the face of change, this understanding is problematic because it implies that change is best avoided. The view developed in this article (and several others in this special issue), is different because it sees change as not only inevitable, but as desirable (Chandler 2021; Korosteleva and Petrova 2021). This article argues that rather than being about *resisting* change, resilience is about *embracing* change through reflective self-governance strategies to continuously adapt to change to not only overcome adversities and continue to be fit for purpose (Schmidt 2015, 406), but also to actually move forward towards a better future. As suggested by David Chandler, this form of resilience-thinking challenges framings of order that prioritise stability and linear conceptions of progress, in doing so opening the possibility that 'the uncertain, the uncontrollable and the unknown can be liberatory rather than oppressive or problematic' (Chandler 2021). It is from this perspective that I see 'Westlessness' as an opportunity for the LIO to 'jump forward' towards adaptation and renewal, rather than as a crisis from which the LIO must recover by 'bouncing back' to a past that has for too long been dominated by 'Westfulness'.

The challenge is that undertaking adaptive behaviour is more easily said than done. While the 'traditional resilience' approach focused on material or systemic characteristics, the 'resilience-thinking' approach in contrast focused on practice and therefore brought human agency and human emotions into play (Hutchison 2019). Once agency and emotions are brought into the equation, resilience is inescapably about agent-based sensitivity to emergent change and agent-led feedback processes to trigger appropriate reaction to change before a disconnect between the purpose and function of the social domain manifests itself (Chandler 2014, 52; Flockhart 2020). This is clearly a complex matter that touches on the psychological disposition of human beings to have a distinct need for cognitive stability. It is well known both from social psychology (Hogg and Abrams 1988) and the literature on ontological security (Laing 1969) that human beings always seek to maintain cognitive consistency through established functional, and often habitual, practice and through striving towards having a stable identity supported by a sense-making narrative that can maintain their biographical continuity (Mitzen 2006; Stele 2008). This is a basic human need that is fully recognised in change management studies (Brännmark and Benn 2012) which have demonstrated that most change processes fail because people tend to actively resist change as they hang on to habits and routines as a bulwark against threatening anxieties (Giddens 1991, 39).

Despite humankind's clear inclination to avoid change, some forms of change seem easier to undertake than others. Changes that can be undertaken without disrupting the biographical continuity of a person or community appear to be, unsurprisingly, easier to undertake than changes that disrupt the deeper social fabric of the community or that affect the identity of an individual. To fully understand how resilience is achieved and maintained, it is therefore necessary to be mindful of different forms of adaptation and renewal. Emanuel Adler (2019) describes the 'sweet spot' between continuous reflective adaption and the preservation of the values that define the essence of a community as 'meta-stability'. Importantly, 'meta-stability' allows for fluctuations in the practices, norms, and rules of a community as it not only adapts to a changing external environment but also keeps its deeper ideational social fabric relatively fixed. From this perspective, a resilient community is one that is able to continuously change its norms, rules and practices in incremental ways as part of a collective learning process whilst maintaining 'meta-stability' through a stable shared understanding of the values that underpin the essential social fabric of the community. To judge if a community such as the LIO is resilient, we must therefore ask what constitutes its deeper social fabric and what it means for it to remain 'fit for purpose'. In the case of the liberal international order, however, doing so is surprisingly difficult.

The term 'liberal international order' is referred to not only in popular and policy debates but also in scholarly debates to denote relatively stable patterns of relations and practices in world politics under American leadership that have facilitated open trade, cooperative security, and multilateralism through the UN system and other multilateral institutions (Ikenberry 2020, 1). Although the current LIO was established in the aftermath of World War Two, it is a continuation of previous forms of liberal orders, most notably the British-led *Pax Britannica* (Dunne and Flockhart 2013). The problem is that there has been considerable tensions, contradictions, disruptions and ambiguities surrounding the purpose, development and values both in the current LIO and in previous forms of liberal orders (Rae and Reus-Smit 2013). However, the tensions, contradictions, disruptions, and ambiguities of the LIO have been successfully hidden in a highly successful and prevalent narrative about the origins, development, and gradual expansion of a specific form of ordering that originated in 16th century Europe. The narrative is prevalent in much of traditional linear history (Hobson 2002) and in liberal accounts of International Relations (Puchala 2003), which present a progressive and Euro-centric account of the gradual expansion of liberal forms of ordering. In particular, the so-called 'expansion narrative' (Bull and Watson 1984) is told in a way that imbues the development and gradual global reach of West-centric forms of ordering with continuity and purpose, although Bull and Watson (1984) acknowledged that this expansion lacked legitimacy and justice. However, even the impression of continuity and meta-stability is a false impression that papers over considerable tensions and contradictions and which fails to address that there is no stable conception—either historically or currently—of what is 'the liberal' in the LIO (Dunne and Flockhart 2013, 69). How important concepts such as order, justice, equality, democracy—even love and family—are understood today do not convey the same meaning as they did in the past, although such terms are presented in the expansion narrative and other forms of linear history as being essentially stable values with an enduring salience. Yet, a closer look reveals that it is almost impossible to identify enduring conceptions of what might be considered 'the liberal in the liberal international order'. The reality is that past liberal order(s) have been based on values, norms and practices that today are seen as illiberal and shameful such as racism and imperialism, or they lacked values that today are seen as indispensable to the contemporary liberal international order, such as democracy and gender equality (Hutchings 2013; Sylvest 2013). Paradoxically, once it is acknowledged that the expansion narrative has been able to incorporate continuous and wide-ranging adaptations and numerous reversals of positions on key issues such as racism, imperialism, gender equality, democracy and human rights, the task of undertaking adaptation and renewal in the current LIO seem less daunting.

The instability of the values underpinning liberal orders in the past is mirrored in both a plethora of names and an ambiguity about the primary values and purpose of the current order. The name *liberal* international order appears to suggest that the purpose of the order is to either defend or engender liberal values such as personal freedoms, a belief in progress, and a commitment to human rights. However, it should be noted that the term 'liberal international order' was hardly used before G. John Ikenberry brought it to prominence in the 1990s and 2000s in his scholarship about American post-war strategy. Moreover, the current international order also goes by the name of the 'rules-based order', the 'American-led order', *Pax Americana*, the 'multilateral order', the 'democratic world order', or simply the 'post-war order' (Jain and Kroening 2019, 11). This plethora of names indicates more than just academic sloppiness, as each name is an indication of an agenda for a specific purpose. Those who emphasise the openness and universality of the LIO point to its rules-based nature and in doing so fully recognise that 'rules-based' can apply to many forms of constituencies, including authoritarian ones. Those who stress the order's commitment to democracy *de facto* put up very specific criteria for membership of the order, in effect making the LIO a rather exclusive club. Despite the ambiguity—and frankly, disagreement—about the precise purpose of the order and the role played by liberal values, those from outside the West have little doubt that the LIO represents a Western-centric perspective that is not only firmly anchored in Western power, principles, and practice, but which also makes very specific demands on those who want to establish relationships with the LIO or join its ranks. Importantly, non-Western actors also have little doubt that the LIO contains a promise to those who join its ranks of a good life that is said to include freedom, peace, and prosperity.

The good life as 'fit for purpose' and as 'trading zones'

This article focuses on the concept the 'good life', but what is meant by this term? The question has been asked from the Stoics to the Existentialists, without leading to a definitive answer. Although all societies rest on shared ideas about what constitutes the 'good life' that are rooted in morality and a sense of justice (Williams 2011, 1237; Wight 1966), the specific understanding of the 'good life' is dependent on the eye of the beholder and is influenced by culture and whether hedonism or virtue is a motivating factor. My ambition here is neither to engage with Stoic nor Existential philosophy to answer a question about what the 'good life' is, as it probably cannot be answered. I merely suggest that although there is no universally agreed definition of the 'good life', the 'good life' as an abstract concept can add to the understanding of resilience by not only capturing the energy that enables human beings to 'keep carrying on' in the face of challenges that could otherwise leave them paralysed and wrecked with doubt and anxiety, but also expressing the essence of a community. From this perspective, the many small ideas and principles that express the essence of a community can gain prominence in debates about order and purpose. It is no coincidence that a common reaction among the commentators to the January 6th storming of the Capitol was to simply declare that 'this is not who we are', because no matter how vague notions of the 'good life' are— it is what comes closest to summarizing the essence of a society and is thus essential for determining what it takes to be fit for purpose. The 'good life' and being *fit for purpose* are therefore inextricably linked and are essential for understanding the essence of the LIO and the challenges to its resilience and opportunities for its renewal.

Although inspired by Aristotle and his insistence that politics must pay attention to 'goods of the soul' such as love, friendship, fairness, self-esteem, honour, morality, and justice (Cooper 1985), my use of the term 'good life' is more pragmatic than it is philosophical. The inclusion of 'softer' considerations that often inform thinking about the 'good life' is an attempt to bring into our reflection on what it takes to be 'fit for purpose', the practical and emotional considerations that are found at the site where ordering practices take place (Bueger and Gadinger 2017, 61), but which are often left out of analyses because they appear inconsequential and difficult to quantify and categorise. The 'good life' captures what Schatzki (2002) refers to as 'teleo-affective structures' that emphasise normativity through a complex and continuously changing array of emotions and moods. These altogether express understandings of what makes sense to do to realize the community's shared vision for the future (Bueger and Gadinger 2017, 63). I argue in this article that although visions for the good life are infinitely complex, always ambiguous, and open to interpretation, they nevertheless add an important, though by no means clear, extra layer to our understanding of what it means to be 'fit for purpose'. From this perspective, the good life is different from the more widely used concept— norms—which also expresses the deeper social fabric of a community. It is important to stress however that including the vision for the 'good life' is not intended to be a substitute for norms, principles, and values, but instead is intended to represent an additional layer with different qualities. Norms are widely agreed to be 'collective expectations for the proper behaviour of actors with a given identity' (Katzenstein 1996, 5) and serve as societal constraints for norm-deviating behaviour (Björkdahl 2002, 9). Where norms are reified in socially sanctioned practices and formal regulatory rules (Björkdahl 2002, 13), this is not the case for visions of the 'good life'. Although a vision is infused with a community's identity and imbued in its narrative and collective hopes for the future (Berenskoetter 2011), they are rarely articulated in very precise terms and are seldom translated into socially sanctioned rules. Moreover, although visions for the 'good life' express the deep social fabric of a social entity and cater for parts of life that speak to human emotions and visions for the future, visions for the 'good life' appear, like 'teleo-affective structures', to be more flexible and able to change in response to a changing context— whether that is in response to changes in technology, availability of goods or changes in morality—or the realisation of emerging threats and challenges such as climate change or pandemics.

The addition of the 'good life' when defining what constitutes 'fit for purpose', is an important extra layer to our understanding of the resilience of a variety of social domains. This is the case as no community can claim to be resilient unless it is able to remain 'fit for purpose' and being so must necessarily mean that the entity has the capacity to undertake such action that fulfils its purpose and delivers a realistic,—albeit perhaps a distant—prospect for realising its shared vision for the 'good life'. Paradoxically, it is the impossibility of defining what the 'good life' is, the looseness of its associated practices, and the absence of socially sanctionable rules that make it useful for understanding resilience as self-governance. The idea of the 'good life' expresses— granted, in loose and ill-defined ways—the purpose of a community by capturing its social distinctiveness and deeper understanding of what makes it what it is, without demanding adherence to specific expectations of appropriate behaviour. In this sense, the idea of the 'good life' resembles what Peter Galison (1997) has called 'trading zones' or social spaces where dissimilar groups can find common ground whilst simultaneously disagreeing on their general outlook (Bueger and Gadinger 2017, 11). The 'trading zone' concept was developed in science-technology studies as a means for different disciplines with distinct identities and different scientific or cultural languages and practices, to better speak to and cooperate with each other in a shared enterprise of innovation and together 'leap forward' to new discoveries (Galison 1997). A similar process seems to be able to take place in relation to the social practices associated with the expression of a community's sense of the 'good life'.

The 'good life' as a 'trading zone' can facilitate a space for speaking and learning across difference by emphasizing social distinctiveness and similarities across different communities. The idea of the 'good life' is a safe way to express nuanced differences within a shared overall cultural heritage even though in reality, the distinction between, for example, the French and the Belgians in the EU or between the Mohawk and the Tuscarora in the Iroquois League, may be difficult for outsiders to spot. Despite its subtlety, a community's shared conception of the 'good life' is an important identity signifier as it enables both distinctiveness claims and the maintenance of a strong narrative about the community without appearing threatening to others. Although the articulations of the vision for the 'good life' can appear quite inconsequential, light-hearted, and even quaint, they nevertheless convey important distinguishing traits whilst offering a framework for forging links across diversity. For example, the Danish concept 'hygge', which conveys a sense of having a cosy time of harmony and tranquillity usually involving lit candles, is a highly treasured characteristic of Danish society. However, the deeper meaning and the importance of the practices associated with 'hygge' are not easily conveyed to outsiders nor are they articulated in formal rules. Nevertheless, 'hygge-occasions' can be used to reinforce the purpose and idea of the 'good life and at the same time provide a safe social setting and—a 'trading zone'—in which to cut across divisions and to find common ground. A more serious way of using the subtlety of visions of the 'good life' can be glimpsed in the Rwandan idea of 'agaciro' (Rutazibwa 2014), which has been consciously constructed as the essence of being Rwandan and as a 'trading zone' for uniting Hutus and Tutsis across the traumas of the genocide (Rutazibwa 2014, 297). The point is that shared understandings of what makes for a 'good life' consist of important, but often overlooked, identity signifiers which not only are crucial for summarizing 'fit for purpose' but which can also be used as a 'trading zone' to achieve understanding and perhaps even friendship and cooperation across much stronger forms of identity signifiers such as ethnicity and religion.

The constitutive elements of ordering domains

The inclusion of the concept the 'good life' is an important addition to our understanding of what is regarded as the purpose of an ordering domain such as the LIO. At the most basic level, the purpose of an international order is clearly to ensure order. However, this raises the questions of what is meant by order and whose order should prevail. Whilst there is general agreement that order relates to a situation of certainty, characterised by predictable behavior for a particular purpose and in accord with recognised norms (Lebow 2018, 20; Bull 1977), what this purpose is in the LIO is not always clear. Hedley Bull has probably written the most authoritative statement on order in his seminal book *The Anarchical Society* in which he was quite specific that order consists of rules-based patterns of behaviour that uphold the fundamental goals of social life—namely safeguarding of life against violent death, ensuring that promises once made will be kept, and respecting agreements once entered—also called the values of 'life, truth and property' (Bull 1977, 5). Bull was insistent that life, truth and property values represent a universal purpose because all societies have a common interest in ensuring the achievement of these values because without them, there could be no society (Bull 1977, 51). However, it must be acknowledged that securing the elementary goals of social life does not promise a 'good life' and the goals mention nothing about what Aristotle referred to as the 'goods of the soul', which arguably are needed for a life that is more than mere existence. Just as there can be no society without a reasonable expectation of the stability of life, truth, and property, it is also hard to imagine a lasting (resilient) society that lacks a shared vision for the 'good life'. Indeed, it would be a bleak existence if order is simply the fulfilment of the fundamental goals. For life to be more than just existence and for it to be worth living, the condition of order must also include a vision for what makes a 'good life'.

Bull's pioneering work on order is now more than forty years old and has itself undergone considerable adaptation and renewal (Suganami 2017). In particular, the form of English School theorising that implicitly draws on constructivism (Dunne 1995) is a useful addition to the resilience-thinking perspective that is put forward in this Special Issue. The recognition that order is a constructed and socially sanctioned condition anchored in patterns of power and identity, norms and values, and formal and informal institutions (Reus-Smit 1997, 2013, 2017; Buzan 2004; Lebow 2008; Phillips 2011) provides a bridge between the resilience-thinking perspective and the constructivist elements of the English School. The constructivist understanding of how order is forged has built on Bull's rather implicit attention to power and identity as the social structure of international society. It does so by focusing on the hegemonic nature of international orders (Nexon 2009; Zarakol 2017), which stresses that appropriate behaviour within an international society invariably reflects the identity of the most powerful member(s) of the society who also determine which norms and values are deemed to be salient for the society as a whole (Barnett and Finnemore 1999, 710). In addition, the literature on order has developed since the 1970s from viewing institutions mainly as structures that constrain behaviour within certain boundaries of appropriateness, to viewing institutions as practice. The latter view emphasises the relationship between human action and social order (Bueger and Gadinger 2017, 60) and views institutions not only as performative sites (mainly in formal international organisations or regimes) for reifying social norms and organisational procedures but also as sites for reflexivity and undertaking change-making action (Spandler 2015; Friedner-Parrat 2017).

The eclectic theorising that increasingly characterises constructivist-inspired English School thinking is particularly important for the resilience-thinking perspective, as it opens for a more agent-centred focus that sees human beings as intensely reflexive social beings. Humans act according to deep psychological dispositions with the clear expectations that their social group (their community or the ordering domain) not only can ensure their physical survival, legal rights and possession of property, but also can enhance their social standing (Lebow 2008), social distinctiveness (Forsby 2015), self-esteem (Rubin and Hewstone 1998), and ontological security (Giddens 1991; Flockhart 2016). The fulfilment of agents' expectations requires a continuous process of narrative construction both to express the identity of the community in question and to ensure alignment between identity and the narrative (Ciutâ 2007) to establish a sense-making link between the past and the present and visions for the future. Narratives are essential for maintaining the social cohesion of a society, ensuring convergence around its history, purpose, and achievements, and facilitating a sense of continuity and order (Ezzy 1998, 241). A strong narrative is an essential condition for agents' experience of their society as ordered and cohesive, and presents the essence of the society such that it outwardly displays its social distinctiveness in a way that enhances the status, self-esteem and ontological security of the collective entity and its individual members (Flockhart 2012, 80).

To be able to more clearly pinpoint where an ordering domain needs adaptation or renewal, and to more precisely identify which aspects of an ordering domain are contested, it can be helpful to focus on the order's individual constitutive elements by moving up the ladder of abstraction to the level of ideal types. Doing so invokes the above understanding of order as being constituted in social domains (communities) where practices of ordering take place. Ideal-type ordering domains can be conceptualised as consisting of three constitutive elements. First is **power**, especially the *mode of power* such as hegemony (Buzan and Lawson 2015) which is always manifested in a specific social identity largely defined by the hegemon (Zarakol 2017). Second are **principles** which are derived from power and identity patterns and specifies both the norms and rules that define appropriate behaviour and social sanctions within the social domain. Third are **practices** (Adler and Pouliot 2011; Bueger and Gadinger 2017) that are performed either through evolved and largely habitual forms of practice that take on a structural quality by reinforcing extant patterns of power and principles, or through intentional and reflexive action that can either undertake small changes and adaptation to maintain the ordering domain's meta-stability, or, more rarely, undertake transformative change intended to bring about renewal. Importantly, the three constitutive elements converge around more loosely defined, but nevertheless important, shared ideas about what constitutes the good life. To assess if such an ordering domain is 'fit for purpose', agents working on behalf of the ordering domain must search for inconsistencies and misalignments within and between the

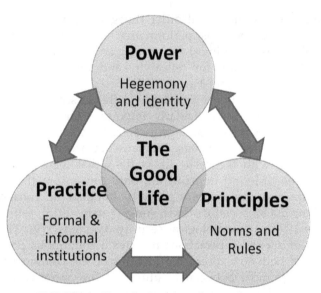

FIGURE 1. The ideal-type ordering domain.

three constitutive elements, asking if they together reinforce the normative purpose of the ordering domain expressed in its vision for the 'good life'. If no, or only few, inconsistencies are found, the ordering domain is 'fit for purpose' and no further action is required for the time being. However, in a dynamic and challenging environment such as the current one, inconsistencies and misalignments are likely to emerge continuously. In such a case, a more thorough uncovering of the specific misalignments needs to be undertaken and followed up by action to undertake relevant adaptation or renewal.

The three constitutive elements are inextricably linked and must be able to be narrated into a sense-making narrative in which the identity specified by the hegemon is clearly reflected in the practices and principles of the ordering domain. This means that whenever change occurs in one of the constitutive elements, either as a response to external stimuli or as a result of agent-based reflection and/or agent-based processes to maintain a stable identity and a strong narrative (Flockhart 2016), it will inevitably require adaptation in the two other constitutive elements and sometimes in the conception of the 'good life'. The condition of order cannot, therefore, be understood without reference to the ordering domain's vision for the 'good life'. This relationship is expressed graphically in Figure 1.

An ordering domain such as the LIO is resilient when agents working on its behalf are able to continuously scan the order's constitutive elements for inconsistencies, contestations and misalignments *and* when those agents are able and willing to undertake the necessary adaptive action vis-à-vis its patterns of power, principles and practice (Flockhart 2020). These processes will likely take place continuously in small and incremental ways as part of a collective learning process within the boundaries of the prevailing shared understanding of the values that both underpin its social fabric and define its purpose and vision for the future (Berenskoetter 2011; Korosteleva and Petrova 2021).

Towards renewal of the liberal international order

Today it is widely recognised that the LIO is in crisis. The concerns voiced in the Munich Security Conference Report and perceptively presented as 'Westlessness' (Bunde et al. 2020), are clearly vindicated when attention is focused on each of the constitutive elements of the current LIO. Although a full and detailed analysis of the challenges faced by each of the constitutive elements is not possible within this article, it seems clear that each constitutive element is facing either crisis, contestation or fading salience and that there currently is a substantial misalignment between the three constitutive elements. Moreover, the sense of 'good life' is in flux with further misalignment to be expected.

It is already well-recognised that the power patterns of the LIO are changing globally through the relative decline of the United States and through contestations against American hegemony by emerging and rising powers. At the same time, the United States is undergoing a form of identity crisis reflected in for example the Black Lives Matter movement, self-reflections following the storming of the Capitol, and in current struggles within the Republican Party. Although these are primarily domestic matters, they are nevertheless important for the LIO because they affect the overall identity and resilience of LIO. Domestic struggles in the leading state are not a small matter and will likely reverberate through all constitutive elements of the order, in effect requiring substantial realignment not only in the other constitutive elements but also in the narrative about what is conceived as the 'good life' in the LIO.

The other important point to note is the surge of contestations against the value base of the liberal order, which directly affects the 'principles element'. It is important to note that current contestations are not just expressions of passive disapproval about abstract ideas but represent legitimate concerns about down-to-earth issues such as the damaging consequences of unfettered globalisation, looming planetary destruction, mounting costs associated with the 'forever wars', and the unequal distribution of the negative consequences of neoliberal policies. In this current situation, the always lurking right-wing populism and nationalism that have lately surfaced in both new and established democracies (including the United States) shows not only that democracy is a fragile condition but also that illiberal forms of democracy appear to have a broad populist appeal. Madeleine Albright's (2018) book *Fascism—A Warning* expresses a growing awareness that the survival of democracy cannot be taken for granted because in the words of Kagan, 'the people in a democracy, excited, angry and unconstrained, might run roughshod over even the institutions created to preserve their freedoms' (Kagan 2016). The point here is that values such as democracy previously thought to be 'chiselled in stone' in the liberal order's foundations might not in the future be part of the deep social fabric of the order. If so, then none of the assumed essential values are in fact safe and thus casts doubt about not only how the resilience of the liberal order can be maintained but also how contestations can be met. This is the case because there is no longer a solid shared sense of what constitutes the 'good life' and what 'fit for purpose' means.

Today there is a growing disconnect between the order's narrative and the perceptions of the 'good life' among those who ideally would like to develop closer relationships with the LIO. The problem is that although the LIO (at least rhetorically) has celebrated cultural diversity, in practice its celebration has been selective and has ineffectively utilised visions for the 'good life' as a 'trading zone' that can establish a degree of unity across diversity. In practice, the LIO has pursued a norm-and-value-based course of action that has implicitly insisted on members adhering to a Western conception of the 'good life'. This arguably has diminished the LIO's legitimacy and salience among both Western and non-Western constituencies (Cooley and Nexon 2020). In this Special Issue where the connectivity and complexity of the modern world is reflected in the attention paid to the importance of the many forms of resilience as self-governance that co-exist between the local and global levels (Korosteleva and Flockhart 2020; Korosteleva and Petrova 2021), the question of whose order and whose conception of the 'good life' matter is particularly pertinent. The editors and several contributors to this special issue are concerned that considerations about whose order, whose resilience, or whose 'good life' matter will continue to be geared towards a Western or Euro-centric application of resilience-building. As pointed out by Korosteleva and Petrova (2021, this issue), the transfer of aid or granting of privilege in concrete policy-areas is still conditional on the receptor community adopting the complete package of Western norms and practices. This is done with little regard for how these Western and Euro-centric norms and practices align with local visions for the good life. Such policies do not bode well for maintaining or extending the appeal of the LIO to communities on its margins.

Realistically, it seems that the time has passed for when it was appropriate to take adaptive action—this was attempted without success with the so-called Princeton Project back in 2006 (Ikenberry and Slaughter 2006). What is now needed is a process of *renewal*, but renewal will involve a major overhaul of both the power, principles and practices of the LIO—and some deep soul-searching on what can be said to be its vision of the 'good life'. The challenge will be to define renewed patterns of power, principles and practice that can better accommodate the visions of both current and prospective members of the order. The ongoing crisis might therefore require more substantial adaptation than most are willing to contemplate. The point is that times have changed, and the sooner policymakers accept this reality, the better. Major shifts are taking place within the LIO and in its external environment that will need to be met with targeted renewal of each of the order's constitutive elements. In the process of seeking to re-establish the resilience of the LIO, a choice between two strategies for restoring resilience exist:

1. To follow a strategy of 'bouncing back' to an adapted liberal order that is more firmly anchored into its traditional liberal credentials, commits to live by traditional liberal values, and seeks to deliver on the implicit promise of the LIO of providing peace, prosperity and freedom to its core members and those associated in a more marginal position.
2. To follow a strategy of 'jumping forward', one which recognises not only the fundamental change of the global and local context but also that the time for mere adaptation has passed. This strategy focuses on renewal of the liberal order's conception of the 'good life' to make it more of a

'trading zone' by focusing on issues such as equality, opportunity, tolerance, and fairness.

The LIO has, in the past, been adept in maintaining its resilience and meta-stability through continuous processes of adaptation within the context of a West-centric narrative about continuity and progress. However, in the new situation of 'Westlessness', there is a need for wholly different processes of reflection on the very essence of the social fabric of the international order. This inevitably will be associated with considerable costs and difficult decisions. The option to 'stay robust in the face of change' by taking no action or anchoring into what is perceived to be the core liberal values is what is entailed in the 'bouncing back' strategy, which seems to be the politically preferred strategy of the Biden Administration and many liberal internationalist. However, this strategy does not address the issue of 'whose good life' will matter and will likely lead to a reduction in the attractiveness of the LIO to potential members and perhaps even to a loss of some current members. Moreover, if only gentle adaptive action is taken, the current crises are likely to continue and might eventually lead to the collapse of the order as failure to undertake renewal is bound to have further negative consequences in each of the order's constitutive elements. The further damage could include damage to the legitimacy of the **power** patterns and uncertainty about the identity of the order; fading salience of the **principles** of the order and increasing instances of rule-breaking or failure to abide by the order's established norms; and a prevalence of habitual **practice** with little energy or enthusiasm for taking action to meet the many challenges facing the global international society. Failure to take the necessary action will likely lead to brittleness of the institutions of the order, perceptions of inefficiency, and withdrawal of support within the wider population. In other words, the politically palatable options of doing nothing or modestly adapting by anchoring into the existing value base will on one hand likely lead to a smaller but a more liberal order. On the other hand, it could lead to the eventual collapse of an order that tries to maintain its narrative of openness and tolerance but becomes increasingly shunned by those who perceive that their conceptions of the 'good life' and their social distinctiveness have not be catered for.

The other option to 'jump forward' through extensive renewal of the LIO also carries significant risks. As several of the articles in this Special Issue indicate, the preservation of local culture, customs and visions of the 'good life' in the wider Eurasia cannot be compromised by outside-in forms of local resilience-building that rest on a West-centric perception of the 'good life'. Therefore, the renewal of the LIO will need to go a considerable way towards a new shared understanding of the 'good life' and what 'fit for purpose' means. If the second strategy of 'jumping forward' is to be adopted, the implications would be significant in practiced policy. It will necessitate sensitivity and respect for the cultural distinctiveness of associated communities with the understanding that all promoted norms, values or practices must cohere with the new vision for the 'good life' within local communities. There is a growing sensitivity to local knowledge and cultural distinctiveness, and global aspirations for inclusivity, tolerance and learning. However, thus far, the possibilities afforded by utilising the concept of the 'good life' as a trading zone for learning and common understanding between different communities and the institutions of the LIO seem distant. The political palatability of any significant change to what is perceived to be core liberal orders is not yet at a stage where such a course of action is realistic. The likelihood is therefore that the crises of 'Westlessness' will remain a concern within the West, whereas contestations against 'Westfulness' will continue outside of the liberal international order.

Acknowledgements

I wish to thank Elena Korosteleva and Irina Petrova for their comments and support in the development of this article. Also thanks to the three anonymous reviewers for their comments on earlier versions of this article.

Disclosure statement

No potential conflict of interest was reported by the author.

Funding

This work was supported by the GCRF-funded COMPASS project 'Comprehensive Capacity-Building in the Eastern Neighbourhood and Central Asia: research integration, impact governance and sustainable communities' (ES/P010849/1).

ORCID

Trine Flockhart ⓘ http://orcid.org/0000-0002-3709-2493

References

Adler, Emanuel. 2019. *World Ordering: A Social Theory of Cognitive Evolution*. Cambridge: Cambridge University Press.
Adler, Emanuel and Vincent Pouliot. 2011. *International Practices*. Cambridge: Cambridge University Press.
Albright, Madeleine. 2018. *Fascism—A Warning*. New York, NY: Harper Collins.
Barnett, Michael and Martha Finnemore. 1999. "The Politics, Power and Pathologies of International Organizations." *International Organization* 53(4): 699–732.
Berenskoetter, Felix. 2011. "Reclaiming the Vision Thing: Constructivists as Students of the Future." *International Studies Quarterly*. 55(3): 647–668.
Björkdahl, Annika. 2002. "Norms in International Relations: Some Conceptual and Methodological Reflections." *Cambridge Review of International Affairs* 15(1): 9–23.

Brännmark, Mikael and Suzann Benn. 2012. "A Proposed Model for Evaluating the Sustainability of Continuous Change Programmes." *Journal of Change Management* 12(2): 231–245.

Bueger, Christian and Frank Gadinger. 2017. *International Practice Theory*. Houndsmills: Palgrave

Bull, Hedley. 1977/1995. *The Anarchical Society: A Study of Order in World Politics*. Houndsmills: Macmillan.

Bull, Hedley and Adam Watson, eds. 1984. *The Expansion of International Society*. Oxford: Oxford University Press.

Bunde, Tobias, Randolf Carr, Sophie Eisentraut, Christoph Erber, Julia Hammelehle, Laura Hartmann, Juliane Kabus, Franziska Stärk, and Julian Voje. February 2020. *Munich Security Report 2020: Westlessness*. Munich Security Conference, Munich. doi.org/10.47342/IAQX5691.

Buzan, Barry. 2004. *From International to World Society?* Cambridge: Cambridge University Press.

Buzan, Barry and George Lawson. 2015. *The Global Transformation—History, Modernity and the Making of International Relations*. Cambridge: Cambridge University Press.

Chandler, David. 2014. "Beyond Neo-liberalism: Resilience, the New Art of Governing Complexity." *Resilience*. 2(1): 47–63.

Chandler, David. 2021. "Becoming Resilient: The 'Right to Opacity' as the Foundation of 'Relations of Community'". *Cambridge Review of International Affairs*.

Ciutâ, Felix. 2007. "Narratives of Security: Strategy and identity in the European Context". In *Discursive Constructions of Identity in European Politics*, edited by Richard C.M. Mole, 190–207. Houndsmills: Palgrave.

Cooley, Alexander and Daniel Nexon. 2020. *Exit from Hegemony: The Unraveling of the American Global Order*. Oxford: Oxford University Press.

Cooper, John M. 1985. Aristotle on the Goods of Fortune. *The Philosophical Review*. 94(2): 173–196.

Dunne, Tim. 1995. "The Social Construction of International Society." *European Journal of International Relation*. 1(3):367–389.

Dunne, Tim and Trine Flockhart. eds. 2013. *Liberal World Orders*. Oxford: Oxford University Press.

Dunne, Tim and Christian Reus-Smit, eds. 2017. *The Globalization of International Society*. Oxford: Oxford University Press.

Ezzy, D. 1998. "Theorizing Narrative Identity: Symbolic Interactionism and Hermeneutics." *The Sociological Quarterly* 39(2): 239–252.

Flockhart, Trine. 2012. "Towards a Strong NATO Narrative: From a 'Practice of Talking' to a 'Practice of Doing'." *International Politics* 49(1):78–97.

Flockhart, Trine. 2016. "The Problem of Change in Constructivist Theory: Ontological Security Seeking and Agent Motivation." *Review of International Studies* 42(5): 799–820.

Flockhart, Trine. 2020. "Is This the End? Resilience, Ontological Security, and the Crisis of the Liberal International Order." *Contemporary Security Policy*. 41(2): 215–240.

Forsby, Andreas. 2015. "The Logic of Social Identity in IR: China's Identity and Grand Strategy in the 21st Century." PhD Thesis, Department of Political Science, University of Copenhagen, Denmark.

Friedner-Parrat, Charlotta. 2017. "On the Evolution of the Primary Institutions of International Society." *International Studies Quarterly* 61(3): 623–630. https://doi.org/10.1093/isq/sqx039

Galison, Peter. 1997. *Image & Logic: A Material Culture of Microphysics*. Chicago: The University of Chicago Press.

Giddens, Anthony. 1991. *Modernity and Self-Identity: Self and Society in the Late Modern Age*. Cambridge: Polity.

Hobson, John M. 2002. "What's at Stake in "Bringing Historical Sociology Back into International Relations?" Transcending "Chronofetishism" and "Tempocentrism" in international relations." In *Historical Sociology of International Relations*, edited by Stephen Hobden and John M. Hobson, 3–41. Cambridge: Cambridge University Press.

Hogg, Michael and Dominic Abrams. 1988. *Social Identifications: A Social Phycology of Intergroup Relations and Group Proce*sses. London: Routledge.

Holling, C. S. 1973. "Resilience and Stability of Ecological Systems." *Annual Review of Ecology and Systematics* 4(1): 1–23.

Hutchings, Kimberly. 2013. "Liberal Quotidian Practices of World Ordering." In *Liberal World Orders*, edited by Tim Dunne and Trine Flockhart, 157–172. Oxford: Oxford University Press.

Hutchison, Emma. 2018. "Emotions, Bodies, and the Un/Making of International Relations." *Millennium* 47(2): 284–298.

Ikenberry, John G. 2001. *After Victory*. Princeton: Princeton University Press.

Ikenberry, John G. and Annemarie Slaughter. 2006. "Forging a World of Liberty Under Law: US National Security in the 21st Century." *Final Report of the Princeton Project*. Princeton NJ: The Woodrow Wilson School of Public and International Affairs.

Jain, Ash and Matthew Kroening. 2019. *Present at the Creation: A Global Strategy for Revitalizing, Adapting and Defending a Rules-Based International System*. Washington, DC: Atlantic Council of The United States.

Kagan, Robert. 2016. "This is how fascism comes to America." *Washington Post*, 18 May. https://www.washingtonpost.com/opinions/this-is-how-fascism-comes-to-america/2016/05/17/c4e32c58-1c47-11e6-8c7b-6931e66333e7_story.html

Katzenstein, Peter. 1996. *The Culture of National Security: Norms and Identity in World Politics*. New York, NY: Columbia University Press.

Korosteleva, Elena A. 2020. "Reclaiming Resilience Back: A Local Turn in EU External Governance." *Contemporary Security Policy* 41(2): 241–62.

Korosteleva, Elena A. and Trine Flockhart. 2020. "Resilience in EU and International Institutions: Redefining Local Ownership in a New Global Governance Agenda." *Contemporary Security Policy* 41(2): 153–175.

Korosteleva, Elena A. and Irina Petrova. 2021. "From 'the Local' to 'the Global': What Makes Communities Resilient in Times of Complexity and Change?" *Cambridge Review of International Affairs*.

Laing, Robert D. 1969/1990. *The Divided Self: An Existential Study in Sanity and Madness*. New York, NY: Penguin.

Lebow, Richard N. 2008. *A Cultural Theory of International Relation*s. Cambridge: Cambridge University Press.

Lebow, Richard N. 2018. *The Rise and Fall of Political Ord*ers. Cambridge: Cambridge University Press.

Mitzen J. 2006. Ontological Security in World Politics: State Identity and the Security Dilemma. *European Journal of International Relations* 12(3): 341–370.

Nexon, Daniel. 2009. *The Struggle for Power in Early Modern Europe: Religious Conflict, Dynastic Empires, and International Cha*nge. Princeton: Princeton University Press.

Phillips, Andrew. 2011. *War, Religion and Empire: Transformation of International Orders*. Cambridge: Cambridge University Press.

Puchala, Donald, J. 2003. *Theory and History in International Relations*. London: Routledge.

Rae, Heather and Christian Reus-Smit. 2013. "Chapter 4: Grand Days, Dark Palaces: The Contradictions of Liberal Ordering Practices. Ordering." In *Liberal World Orders*, edited by Tim Dunne and Trine Flockhart, 87–105. Oxford: Oxford University Press.

Reus-Smit, Christian. 1997. "The Constitutional Structure of International Society and the Nature of Fundamental Institutions." *International Organizatio*n 51(4): 555–589.

Reus-Smit, Christian. 2013. "The Liberal Order Reconsidered." In *After Liberalism?: The Future of Liberalism in International Relations*, edited by R. Friedman, K. Oskanian and R. P. Pardo, 167–186. Basingstoke: Palgrave Macmillan.

Reus-Smit, Christian. 2017. "Cultural Diversity and International Order." *International Organization* 71(4): 851–885.

Rubin, M. and M. Hewstone. 1998. "Social Identity Theory's Self-Esteem Hypothesis: A Review and Some Suggestions for Clarification." *Personality and Social Psychology Review* 2(1): 40–62.

Rutazibwa, O.U. 2014. "Studying Agaciro: Moving Beyond Wilsonian Interventionist Knowledge Production on Rwanda." *Journal of Intervention and Statebuilding* 8(4): 291–302.

Schatzki, Theodore. 2002. *The Site of the Social—A Philosophical Account of the Constitution of Social Life and Change*. Philadelphia: Penn State University Press.

Schmidt, Jessica. 2015. "Intuitively Neoliberal? Towards a Critical Understanding of Resilience Governance." *European Journal of International Relations* 21(2):402–426.

Spandler, Killian. 2015. "The Political International Society: Change in Primary and Secondary Institutions." *Review of International Studies* 41(3): 601–622.

Stele, Brent. 2008. *Ontological Security in International Relations Self-Identity and the IR State*. London: Routledge.

Suganami, Hidemi. 2017. "Chapter 2: The Argument of *The Anarchical Society*." In *The Anarchical Society at 40*, edited by Hidemi Suganami, Madeline Carr and Adam Humphreys, 23–40. Oxford: Oxford University Press.

Sylvest, Casper. 2013. "Chapter 9: Theoretical Foundations of Liberal Order." In *Liberal World Orders*, edited by Tim Dunne and Trine Flockhart, 173–192. Oxford: Oxford University Press.

Wiener, Antje. 2017. *A Theory of Contestation*. Berlin: Springer.

Wight, Martin. 1966. "Why is there No International Theory?" In *Diplomatic Investigations*, edited by Herbert Butterfield and Martin Wight. 33. London: Allen & Unwin.

Williams, John. 2011. "Structure, Norms and Normative Theory in a Re-defined English School: Accepting Buzan's Challenge." *Review of International Studies*. 37(3): 1235–1253.

Zarakol, Ayse. 2017. "Theorising Hierarchies". In *Hierarchies in World Politics* edited by Ayse Zarakol, 1–14. Cambridge: Cambridge University Press.

Encountering the Complexity of Global Life

Emilian Kavalski ⓘ

Abstract *This chapter lays foundation for rethinking our understanding of the world we live in today, from a novel Complexity Thinking perspective. The chapter challenges the conventional premises of International Relations, to suggest that it is the emergence, uncertainty, non-linearity and insecurity that shape the world, as well as human, and non-human interactions within it. Essentially, what makes and explains the world of today is its interwovenness, and relationality; and to this end, everything 'global' has its roots in 'the local'; and life as we know it, is situated in, and driven by, the often-hidden relations of the 'inside/outside' and the 'around' environment. The chapter explores the nuances of this complexity-thinking through the logics of (in)security ('inside/outside') and resilience ('around') as a premise for imagining and explaining the emergent world, where community of relations (be it human or non-) remains central to its interactions.*

Introduction

The late scholar of world affairs James N. Rosenau (1960, 21) once observed that the student of world politics 'embarks on a search for certainty, only to find that it lies in such phrases as "apparently", "presumably", and "it would seem as if".' Thus, 'apparently', to use Rosenau's suggestion, uncertainty has always been a defining feature of the study and practice of world affairs. So why then are policy-makers, International Relations (IR) scholars, and publics surprised when the world turns out to be unpredictable? After all, depending on how far back one is willing to look, the discipline (at least in its 'Eurocentric' form) has been around for quite some time since the first department of international politics opened its doors at Aberystwyth or since Thucydides scripted his account of the Peloponnesian wars. In either case, the veritable age of IR should have 'presumably' provided it with enough experience to expect—if not necessarily be prepared for—the unexpected. Yet, as Rosenau (1980) reminds us, IR is anything but prepared for uncertainty (and has been so for a while). According to him, 'it would seem as if' the mainstream has lost its 'playfulness'. Thus, instead of allowing 'one's mind to run freely, to be playful, to toy around with what might seem absurd, to posit seemingly unrealistic circumstances and speculate what would follow if they ever were to come to pass', the IR mainstream has sidelined its mischievous nature in favour of stiff parsimonious models simplifying the contingent nature of most that passes in world affairs. For Rosenau therefore it is no wonder that IR has consistently failed to 'imagine the unimaginable' (Rosenau 1980, 19–31).

It is for this reason that he pioneered non-linear approaches 'to toy around' with the complex patterns of world politics (Rosenau 1990). Subsequently, the propagation of Complexity Thinking (CT) concepts and ideas across the IR domain has become one of the most fascinating – if overlooked – trends in the discipline. Complex challenges emerging from the interconnectedness between local and transnational realities, between financial markets and population movements, and between pandemics, a looming energy crisis, and climate change have tested IR's ability to address convincingly their turbulent dynamics. Such complex challenges intimate a pattern of interactions marked by sharp discontinuities, which make modern, large-scale actors – such as states and international organizations – vulnerable to unexpected shocks. The emergence of such qualitative uncertainties demands a different type of thought process capable of addressing the multitude of forces and random processes that animate the dynamism of global life (Bernstein et al. 2000).

The contention is that the discipline has increasingly immersed itself in debates on the substantiation of particular paradigms rather than engaging with the reality of global, and as complexity-thinking insists, of local life. To put it bluntly, the turbulence of world affairs appears to have relevance (primarily) to the extent that it can validate (or disprove) the proposition of a particular IR school, including to a degree of vilifying everything 'local' as irrelevant or inconsequential to global affairs. Some have even called this a 'castle syndrome' – namely, a disciplinary condition in which proponents of different IR schools engage in defending and reinforcing the bulwarks of their own analytical castles, while bombarding the claims of everybody else (Barkin 2010). Such contention should not be misunderstood as a condemnation of the field, or as a suggestion that it lacks sophistication. The suggestion here is that despite the 'new challenges', IR has not abandoned its 'old habits' (Waltz 2002). Such a proclivity has recently been termed as 'returnism' – IR's predilection for traditional conceptual (global) signposts that provide intellectual comfort zones, but are 'simply images of old concepts' de-contextualized from (and, therefore, inapplicable to) current realities (Heng 2010).

Such a mentality has hindered the interaction between the different IR paradigms, between IR and the advances in other social and natural sciences, as well as the development of qualitatively new intellectual platforms for engaging the complexity of world affairs. The suggestion of this chapter is that while IR scholars often employ the metaphor of complexity, relationality or resilience, the potential theoretical and policy contributions emerging from the analytical principles of CT have largely been neglected. The marginalization of CT proponents within the discipline reflects both their refusal to think-in-paradigms and the espousal of a new vocabulary both for the study of IR and for the explanation and understanding of global life as essentially 'local' and 'relational', which very often has its origins in the natural rather than the social sciences.

However, one question that needs to be addressed at the outset is: Why complexity thinking? The answer offered by the proponents of CT is that IR needs new forms of knowledge to respond to emerging complex challenges; in particular knowledge coming from a different epistemological, ontological, and ethical place than the conventional repertoire of IR (Murphy 2000). CT offers such a point of departure. In particular, CT endeavours a form of argument which illuminates that the development of sophisticated and sustainable responses to current challenges requires the recognition of complexity—not for complexity's own sake, but because simplistic solutions are unsustainable and counterproductive (Kavalski 2012b). What IR can gain from such a move are useful analytical and policy-making concepts and ways of thinking about the dynamism of a fragile and unpredictable global life.

The use of the notion of 'global life' with reference to the local is not coincidental here. It allows the exploration of the full spectrum of CT's contributions to IR. As it will soon become apparent, CT has a (potentially) transformative impact both on the established anthropocentric IR and on the emerging non-anthropocentric one. Having its roots in the Latin word *complexus*—describing 'that which is woven together' as well as something that has 'embraced', 'plaited' several elements—the complexity perspective infers the *interwovenness* of life, be it global or local (both as an inherent quality and a systemic condition). The recognition of such interwovenness between human and natural systems defines *global life* not merely as international politics, but as coexistent 'worlds', 'domains', 'projects', or 'texts' of ongoing and overlapping interconnections, situated locally, within a given time and context (Rosenau 1988). Global life thus consists of more than just political communities and the countries that they form—i.e., it is not only about what happens '*inside/outside*' the state, but also about what happens '*around*' the state, and more essentially, what happens around and about communities that form them. It also reveals that the 'international system' is embedded within wider structural conditions and interactions located within the environment 'around' the conventional focus on inter-state relations—an environment, which conceptually constitutes as well as causally conditions (although not in a mono-causal and linear fashion) states *and* communities as other actors (Kurki 2008, 255–261).

It has to be stated from the outset that such engagement with the 'around' of global/local life is much less radical than it might appear at first sight. In fact, it merely recollects the central place that the agency of nature used to be accorded in the study of IR. For instance, already in the 1920s, the discipline has acknowledged that the natural environment is one of the key actors on the international stage. As Raymond Garfield Gettell insisted, despite 'man's best efforts to bring the world in which he lives under his control, the influence of the natural environment upon political evolution has been throughout all human history an important and, in many instances a decisive, factor... Battles, upon whose outcome the fate of nations has depended, have been decided by natural phenomena such as wind, rain, fog or snow, beyond human control' (Gettell 1922, 322). In particular, the significance of the 'around' of global life to the study and practice of international affairs has been stressed by the suggestion that 'the dominant factor which determines the survival of a group is suitability to the environment' (Heath 1919, 143), this way implicating essentially the local conditions.

In this sense, already from its outset, IR has acknowledged that nature's agency—even if unintentional—plays an important role in the unfolding of world affairs (and should therefore not be discarded). For instance, it is often overlooked that with his emphasis on the 'geographic causation [behind] the competing forces in current international politics' Halford Mackinder, the father of geopolitics, intended not only to draw attention to the crucial role played by geography, but rather 'to exhibit human history as part of the life of the world organism' (a statement, which can be read as Mackinder's version of the notion of the 'around' of global life grounded in the local). From this point of view, the dynamics of world affairs demonstrate that 'man and not nature initiates, but nature in large measure controls [the outcomes]' (Mackinder 1904, 422). Consequently, such recognition of and confrontation with the 'around' of global life calls for

> a major revision of our understanding of international relations: politics among and above nations is recognised as a part of a vast natural system, a biosystem. Therefore, all past units we [have] become accustomed to—territorial units and functional relationship—are subsumed under the biosystemic perspective. All units and all relationships become relevant.
>
> (Haas 1975, 842)

Thus, the emphasis on the notion of global life intends to resuscitate IR's interest in the ossified knowledge about the embeddedness of world affairs in the 'around' that provides the context for what has and makes possible its interactions. Human societies and their international interactions are just 'one component in a package of interdependent life forms that continue to adapt to each other' (Clark 2000, 4). The suggestion is that the 'inside/outside' and the 'around' aspects of the study of world politics are not in contradiction but part of the same spectrum of dynamics embedded in the patterns of global/local life (Cudworth and Hobden 2011; Kavalski 2021). The notion of global life therefore elicits that all human interactions are embedded in and made possible by complex interconnections, connecting everything local to the global and back, and this way creating a complex system – a mesh of relations, that requires a granular view on the relational dynamics. The claim is that in contrast to the conventional distinction between subjects (humans) and the objects (the world around them), the emphasis on the concomitance of the 'inside/outside and around' allows for acknowledging the agency and subjectivity of human and non-human actors on essentially the global/local stage.

The reference to global life should not be misunderstood as an insistence on the similarity of human and non-human systems (be they biophysical or technological). On the contrary, the notion of global life does not deny the qualitative differences between human and non-human systems. Instead, it underscores that the two are mutually implicated and interdependent. In other words, the emphasis on the global life proffers a 'human-in-ecosystem' perspective on the study and practice of IR, which recognizes 'the mutual influence of ecological and social processes, instead of treating social and ecological systems as linked but separate domains' (Davidson-Hunt and Berkes 2003, 54). Thus, while this study does not want to brandish CT as a magic potion to cure the crises plaguing the global condition, it nevertheless suggests that CT offers unique opportunities (if not for blurring the dichotomy between anthropocentric and non-anthropocentric IR) for a thorough reconsideration of the explanation and understanding purveyed by representatives of both anthropocentric and non-anthropocentric IR.

To that end, the exploration undertakes a brief overview of the complexifying trends in IR. The aim is to sketch some of the intellectual provocations that have backstopped CT's contributions to disciplinary inquiry. The study then uses this framework to outline the bifurcated security logics dominating the analytical frameworks of IR's purview—namely, the logic of control and the logic of resilience. The intention is to offer a glimpse into CT's potential to generate new ideas and new arguments for tracking the evolution of global life through periods of discontinuous change, in ways that promise to better over time both understanding and action. The necessary caveat is that such complexification of IR should not be misunderstood as a call for new hegemonic hierarchies privileging one lived reality over another. Instead, CT perspectives do not proscribe antagonism, nor do they suggest that its elimination is required. Difference—including radical difference—is not merely desirable, it is the very condition of possibility for the self-organizing emergence of global life (Pan and Kavalski, 2018).

Complexifying IR

The applications of CT to the study of world politics offer perhaps the best confirmation of the insistence that 'the value of complexity exists in the eye of its beholder' (Manson 2001, 412). As a referent for the intricacy of international processes, 'complexity' has become an integral part of IR discourses as is instanced by the notions of 'complex interdependence' (Nye 1993, 169), 'complex learning' (Wendt 1999, 170), 'complex political emergencies' (Goodhand and Hulme 1999), 'complex security' (Booth 2005, 275), 'complex socialization' (Flockhart 2006), and 'complex political victims' (Bouris 2007)—to name but a few. Yet, despite their sophistication, such uses of the term fall short of suggesting the analytical paradox of the complexity of global life—'the less foreseeable the future, the more is foresight required; the less we understand, the more is insight needed; the fewer the conditions which permit planning, the greater is the necessity to plan' (Ruggie 1975, 136).

In this respect, the proponents of a CT approach to world politics insist that IR scholars are unaware of the built-in limitations of the mainstream agenda (Earnest 2015; Kavalski 2011). In order to address these shortcomings, the application of CT research to IR has cut across the intellectual purview of the discipline. The breadth and scope of this literature corroborates the suggestion of a 'paradigm shift' in the study of world politics (Harrison 2006; Rihani 2002). At the same time, Adler (2005, 32) insists that the application of CT to IR proffers images and sets of perceptions about causality, which are broader and more profound than the concept of 'paradigm' would suggest. Such diversity of views and proponents reveals that there is not one single CT approach to IR, let alone an emergent Complex International Relations theory in the offing; if anything, there is a multitude of contending Complex IR theories, perspectives, and approaches which hint at a nascent complexification of IR.

On a theoretical level, the application of CT to the study of world affairs proffers 'new ways of thinking about how global politics unfold' in an environment where 'uncertainty is the norm and apprehension the mood' (Rosenau 2003, 208). Thus, while most IR scholars would agree that the world of their investigations is complex, they still insist that the proper way for acquiring knowledge about it is through the modelling of linear relationships with homogeneous independent variables that discern between discreet stochastic and systemic effects (Hoffmann and Riley 2002, 308). The value from the complexification of IR is to *start thinking* about the interconnections of global life in terms of complex systems. The application of CT to IR asserts that uncertainty and unanticipated consequences should not be surprising, but should be expected (Beaumont 1994, 155; Cho and Kavalski 2015, 433; Cioffi-Revilla 1998, 25). Although this might seem like a truism, it is surprising how little attention mainstream IR theory spares for the study of contingency and contradictions. In translating the jargon of complexity to the vocabulary of IR, Rosenau (2003, 11) has substituted it with the term 'fragmegration.' His intention is to suggest 'the pervasive interaction between fragmenting and integrating dynamics.' As such, fragmegration

> serves as a constant reminder that the world has moved beyond the condition of being 'post' its predecessor to an era in which the foundations of daily life have settled into new and unique rhythms of their own. Equally important, the fragmegration label captures in a single word the large degree to which these rhythms consist of localizing, decentralizing, or fragmenting dynamics that are interactively and causally linked to globalizing, centralizing, and integrating dynamics.
>
> (Rosenau 2003, 11)

Such accounts seek to challenge the 'engineering mentality' of conventional IR (Kavalski 2016). According to Ken Booth (2005, 37), its mode of analysis illustrates 'essentialist understanding about selfish and fearful humans' and 'reductive theories about "*the* logic of anarchy".' By invoking ontologies of danger and death, the ethos of survival privileges 'the maintenance of human life over and against the world and entities within the world' (Odysseos 2007, 18). Thus, 'the myth of 1648' underscores the central 'technology of control' of conventional IR, which constructs the international as an 'alarming utopia,' which 'believes itself grounded in realism' and 'claims to be founded on the laws of History and an irrefutably scientific prediction' (Morin 2006, 135). In this context, the symbolic meaning of anarchy frames the research agenda of conventional IR as an 'empire of uniformity' driven by the modernization belief in a scientific reordering of the social and natural context (Inayatullah and Blaney 2004, 89). By acknowledging that there are many possible avenues for observing global life, CT seeks to provide a conceptual framework within which IR theory can learn, adapt, and interact 'to maximize its own local interactions and complexity to find its own way' (Geyer 2003a, 254).

The Logics of Security: Inside/Outside and Around

Security has always been one of the most contested and simultaneously one of the defining concepts in the study of international affairs. Its pervasiveness has therefore been compared to oxygen—we become aware of it only when it is absent (Keohane, 2002). While this investigation concurs that there is no overarching logic of security, it nevertheless suggests that the field of IR is populated by distinct grammars of security. This section draws attention to the conventional and complexity-based narratives of security. Although these two logics are presented as opposites, the intention here is not to suggest that they are mutually exclusive (Kavalski 2009). Instead, it is claimed that their cognitive models are bifurcated—in other words, although distinct, their representations are simultaneously parallel, complementary, and contending. Bifurcation posits the conventional and complexity-based logics of security not as either/or alternatives, but as accounts capturing simultaneous, yet ambiguous aspects of the reality of global life.

In fact, Hans Morgenthau's recognition of 'the inevitable gap' between 'the science of political science' and the acknowledgement that the 'political reality' of world affairs 'is replete with contingencies and systematic irregularities' (or what he calls, 'international politics as it actually is') displays the dialectical bifurcation inherent in IR observation (Morgenthau, 1973). Since Morgenthau, however, the study of security in conventional IR has become an 'exercise in forgetting' the bifurcated radical reality of global life (Booth, 2005). Such forgetting seems to permeate the mainstreaming of complex challenges—such as pandemics, environmental crises, or human mobility—in IR. According to Ira Chernus (1993), the framing of complex emergencies confirms IR's self-defeating quest for order and stability. Consequently, contemporary forms of security governance have been premised on a cognitive model aimed at the elimination of risks rather than adaptation to their occurrence.

Security-seeking in this context reveals the need to attain safety and avoid harm at any cost. The repudiation of the uncertainties defining the multiplicity of global life has led conventional IR to adopt a mindset of continuities that makes it difficult to address randomness and complex interactions. Furthermore, the preoccupation with risk-elimination obfuscates the fact that human societies inhabit complex spaces and underpins the paradox of (in)security in IR—that is, 'in security we find insecurity,' because when people try to protect themselves and to create a sense of security, they also produce danger, fear and harm (Der Derian, 2006). As it will be explained shortly, in conventional IR security is usually understood as a condition of stability and regularity. Consequently, it is the very predictability (and ultimately entropy) of secure patterns of relations that makes them vulnerable and maladapted. The complexification of IR draws attention to these issues in an attempt to rectify them.

Logic of control: Inside/Outside

The claim here is that the security discourses of conventional IR are underpinned by a distinct set of perceptions that can be best described through the logic of control. More specifically, this (primarily, national) security narrative reflects the human control over natural space and future temporalities. Such logic of control also evinces a conviction in reversible agency—that is, the effects of actions can be undone (more often than not through technological innovation). Thus, despite assertions by security commentators that the world of their investigations is becoming increasingly complex, they would still downplay discontinuous change in favour of simplifying narratives of tractable linear causality.

The origins of such logic of control are usually traced to the Enlightenment 'faith in a "makeable world"' (Heinzen 2004, 35). It reveals the securing of natural environment for human activity that (over)emphasizes the capacities for conscious influence. Premised on the belief in human rationality and fundamental physical order, the assumption of control implies high levels of predictability by using reductionist methods contending that global life changes in a gradual manner and following foreseeable trajectories. Such logic of security reflects the injunctions of the Newtonian paradigm that (1) cause and effect are proportional; (2) the whole is the sum of its parts; and (3) the measurement/quantification of phenomena can assist with the near perfect prediction of their future trajectories (Clemens 2013; Gunitsky 2013; Kavalski 2012a; Kissane 2011).

The impetus behind such logic of control is both the desire and the belief in the capacity of a system to reduce the anxiety provoked by the confrontation with the complexity of global life. The notion and practices of security are aimed at projecting an orderly and coherent story. Thus, the desire for and the feeling of being *in control* becomes a central aspect of security. It rests on 'command-and-control' strategies that aim at controlling changes in a target community (through citizenship laws, for instance). Thus, the anxiety emanating from the perceived chaos of international anarchy—especially, crises, surprises and sudden and rapid changes—reinforces the hankering after control, which then leads conventional IR to spin into motion an illusory web of the real international politics that obscures (if not altogether banishes) the complexity of global life. Consequently, 'our fear of disorder makes it inevitable that we will either find or create an endless supply of it' (Chernus 1993).

The logic of control thereby depicts security as an attempt at conservation which resists disturbance and change. In this setting, Cynthia Webber and Mark Lacy draw attention to the growing trend of 'security by design' aimed at the production of consumer goods and public spaces by 'designing out insecurity' and 'designing in protection.' By promising a quick technological fix to current insecurities, the agenda of safe living underpinning the practices of security by design presents 'a happy-ever-after portrayal of life where... nothing essential will change, everything will stay the same. The resulting scenarios extend pre-existing reality into the future and so reinforce the status quo rather than challenge it' (Webber and Lacy, 2010). Thus, the logic of control turns security into a synonym for equilibrium. Equilibrium here refers to the lack of learning, change or anticipation driving the maintenance of 'the orderliness of ordinary living' (Johnson, 2002). National life in a sovereign state provides the sense of certainty that the logic of control has to secure.

The logic of control circumscribes the Westphalian state as a closed system, isolated from external threats by its borders, whose *inside* is comprehensively regulated to protect it from the anarchical *outside* of both international affairs and the global environment. The allocation of static subjectivity to states reflects their 'anthropomorphization' in which they assume a number of characteristics usually attributed to human subjectivity such as purposive behaviour, self-help, and rationality (Odysseos, 2007; Zolkos and Kavalski, 2016). The contention here is that owing to this hankering after equilibrium, states have grown unfit for their environment. The logic of control has hindered their co-evolution as a result of the prioritization of stability over change.

Logic of Resilience: Around

The awareness of multiple complex challenges has confronted the assumption that the natural environment provides a passive—both fully knowable and infinitely manipulable—context for human endeavours (Kavalski 2022; also the introduction in this volume). Drawing on their willingness to engage other ontologies as a way of learning different ways to observe and encounter the world, ourselves, and the problems that embroil us, CT-inspired perspectives have situated such alternatives into a nuanced comparative conversation with more familiar critical political lexicons and procedures inherited from academic scholarship. In these settings, relationality discloses the world as a multiversal space where alternative realities can and do coexist and have done so for quite some time (Kavalski, 2021; Kurki 2020). As such, the complexified knowledge-production underpinning the study of global/local life mandates tolerance of at least as much diversity and contradictions as evident in the social/community relations being narrated.

The suggestion here is that security premised on the logic of control contributes to the production of (rather than prevents) current vulnerabilities. In particular, the interlocutors of conventional IR seem to have sought refuge in thinking 'restlessly within familiar frameworks to avoid thought about how [their] thinking is framed' (Connolly, 1993). The attempt to avoid risk has played a crucial role in the dearth of capacity and willingness for adaptation characterizing the mainstream of conventional IR. Instead, the nascent complexified IRs acknowledge the pervasive confrontation of the Eurocentric and anthropocentric agency of everyday life with the 'globalization of contingency'—the appearance of emergencies resistant to human control reproduced through the analytical hegemony of the historically institutionalized patterns of economic growth, mastery over the environment, and territorial sovereignty (Connolly, 1991; Kavalski, 2020b). The recognition of the interwovenness of socio-political and biophysical systems suggests that the requirement for adaptive contingency—namely, 'a capacity to respond to the moment of contingent interaction' (Dillon, 2007).

Global life is not constituted *by, through,* and *in* humans alone. In fact, the endeavours of CT actively seeks to rethink the meaning and practices of being human in the context of the Anthropocene by re-connecting, re-relating, re-worlding socio-political systems to the very complex environments which they inhabit and depend on. The irruptive translation of such coexistence brings in dialogue the form and the content of the languages and experiences of the diverse and infinitely complex worlds cohabiting in global life. Such pluriverse becomes coextensive of and standing together with the interpolating spontaneity of surrounding events and things. In other words, thinking beyond the anthropo-centric frames of IR urges 'us to connect the questions of political possibility with the dynamics and the intransigence of vast domains that are themselves recalci-trant to the purchase of politics' and, at the same time, acts as a provocation 'to imagine worlds both before and after us' (Clark 2000, 27–28).

Such developments have demonstrated that the blurring between the inside/outside of the logic of control because of the 'lack of an "enemy" or "other" to act upon [and] uncertainty about who is doing the acting' (Harris 2009, 33). Instead, security thinking has been confronted with the 'political effects of agents who are not conventionally perceived as "political"'—such as, pandemics and droughts. (Prins 1995; Kavalski 2020a). As a result, the 'threats', 'dangers', and 'insecurity' associated with climate change are not conventionally perceived as intentional—that is, there is no conflict of wills between distinct (and opposing) strategic actors. The poser is how can we defend ourselves meaningfully against adversaries that lack 'agential intentionality', let alone formulate national security policies against entities that can hardly be labelled as 'enemy combatants'? Such queries point to the emergent properties of the interaction between human and non-human systems.

A function of conjunction, emergence intuits that system-wide characteristics do not result from superposition (additive effects of system components), but from interactions among components (Kavalski 2007; Korosteleva 2018; and in this volume). Security in such complexified perspectives tends to be explained and understood through the logic of resilience. Resting on the notion of resili-ence (understood as the elastic capacity of materials to return to their original form after being bent, compresses, or stretched) its percolation into social scien-tific inquiry seems to denote 'the capacity of a system to experience shocks while retaining essentially the same function, structure, feedbacks, and therefore iden-tity' (Walker and Cooper 2011; Korosteleva and Flockhart 2020). The point here is that the resilient 'fix' demands internal capacities for self-organization rather than external imposition or intervention. Such framing suggests more sustain-able 'bottom-up and around solutions' (Petrova and Korosteleva 2021), as well as the need to simply adapt, but transform, with change.

The logic of resilience reflects a shared and communal social capacity to anticipate, resist, absorb, and recover from an adverse or disturbing event or process through adaptive and innovative processes of change, learning, and increased competence (Frerks et al. 2011, 113). This consideration discerns the need for *exaptation* provoked by the mercurial uncertainty, cognitive challenges, and complex unbounded risks characterizing the complexity of global life in the. Exaptation reflects the resilience competences of societies to be adapted (and adaptable) both to present circumstances and to the untimely contingen-cies of the future (Grosz 2004, 11). In security practices, such framing of the logic of resilience indicates an ability to (1) cope successfully with challenging or threatening circumstances; (2) defy destructive pressures; and (3) construct new proficiencies out of unfavourable conditions (Kavalski 2008). Such framing construes risks and threats as *normal* features of global life. As such, insecurity should neither be eliminated nor ignored, but merely acknowledged, managed, and adapted.

Thus, in contrast to the logic of control underpinning the concept of security in conventional IR, the logic of resilience reframes security through the capacity to overcome entropic tendencies through external feedback, and internal restructuring (Alker 2011). Bearing in mind the record of 'security-first' modes-of-thinking, it might not be 'necessarily good to be secure, instead it might be better [to] learn to live with insecurity' (Wæver 1997). Accepting to live in and with insecurity reveals the potential for transcending the dangers of uncertainty, unpredictability, and unexpected change usually associated with it. In the context of global life, such acceptance of insecurity as normal condition of existence (as opposed to something which is exceptional, out of the ordinary, and different) informs new modes of engaging with security analysis. Security analysis thereby functions as a map of complexity, which makes possible the imagining of subjects other than states and interactions other than war. At the same time, the logic of resilience provides a diverse repertoire of responses to 'anticipate the unexpected as the norm' (Fowler, 2008).

Thus, just because an action is irreversible does not mean that it should not be undertaken. Instead, acknowledging this 'ethical complexification,' the logic of resilience prompts an 'ecology of action,' which Edgar Morin calls 'living life'— that is, 'not just living,' but 'knowing how to resist in life' by 'daring the acceptance to risk' (Morin 2004, 43–44; 2006, 143). The logic of resilience, therefore, calls for 'a higher level of reflexivity' and indicates that contingent events bring about opportunities for developing new governance skills and norms (Whiteside 1998, 652). Since political action takes place in a self-organizing context, policy-makers need to accept that their decisions will have unpredictable and (oftentimes) unintended outcomes. The complexity of global life demand intellectual flexibility 'in order to avoid a dogged, single-minded pursuit of an effect that is no longer important or even obtainable in the evolutionary system of strategic interaction… Flexibility requires error, tolerance, and avoidance of over-control' (Sakulich 2001, 38). The resilient capacities implicit in such framework offer the possibility for 'a communicative belonging in an insecure world' (Booth 2005, 268).

Conclusion

How important complexity really is? This is an important question which the blossoming literature with the word *complexity* in its titles seeks to address. Rooted in the conviction that global life outlines a complex mesh of contingent interactions, the complexification of IR draws attention to the ongoing inter-penetration between agency, structure, and order amongst the diversity of power, form, and mater implicated in, enacting, and enabling global and local life (Korosteleva and Petrova 2021). Inhabiting a complex universe reveals not only the interdependence between international actors but also their mutual implication in each other's interactions and roles as well as the overwhelming embeddedness of these relations in the world. According to its proponents, CT offers meaningful and productive ways to negotiate the critical junctures between imperialism, greed, environmental degradation, and hope. Such an endeavour is not intended to brandish CT—both in general and in IR in particular—as either a panacea for the crises plaguing the global condition or the flaws of the disciplinary purview of IR (Cudworth et al. 2018; Kavalski 2016). Instead, the claim is that relational perspectives offer a range of alternative stories that need to be heard.

In this respect, the complexification of IR seeks to uncover modes for understanding, explanation, and encounter not only attuned but also able to sustain complexity, foster dynamism, encourage the cross-pollination of disparate ideas, and engage the plastic and heterogeneous processes that periodically overwhelm, intensify, and infect (while all the time animating) the mercurial trajectories of global life. The interlocutors of CT in IR insist that rather than being fearful of analytical crossroads and the unexpected (and unintended) encounters that they presage, IR should embrace the uncertainty attendant in the journey beyond the substantivist ontology of the world. In contrast to the dualistic bifurcations that dominate IR imaginaries, the encounter and engagement with relationality both illuminates and reminds the study of world affairs that the complex patterns of global life resonate with the fragility, fluidity, and mutuality of global interactions, rather than the static and spatial arrangements implicit in the fetishized currency of self-other/centre-periphery/hegemon-challenger models underpinning the binary metanarratives of IR (Kavalski 2018).

CT simultaneously amplifies and analyses the intrinsic relationality both of global life and the realms of IR. Such complexification uncovers an IR as a project of disclosure—on the one hand, disclosing worlds and possibilities foreclosed by the Eurocentrism and anthropocentrism of its Newtonian bias; and, on the other hand, disclosing the inextricable and invariable intertwinement between understanding, explanation, practices, and encounters in the study of world affairs. Thus, as the patterns of world politics amply demonstrate, in periods of 'rapid, discontinuous, fundamental, global, multicultural change, coherent belief systems are an obstacle to the effective structuring of comprehension and action' (Allenby and Sarewitz 2011, 121). CT's attentiveness to the promise and possibilities of uncertainty makes the realms of IR research doubtlessly messy. Yet, messiness is needed if IR is to recover a disposition for encounter and engagement with currents, trends, and voices occluded by the Newtonian underpinnings of established paradigms. The claim of this study is that the various analytical 'alliances' between CT and IR can allow the study of world politics to develop the skills, frameworks, and governance mechanisms required to 'think the unthinkable' trajectories of global life (Connolly 2011). Such inference echoes Rosenau's intuition that IR has to get comfortable with the power of the contingent and chaotic forces of the fast-changing and complex global life. The key to IR's coping in such dynamic contexts is to be 'able to adjust to the unexpected in creative and appropriate ways' (Rosenau 1970; Rosenau 2001).

Acknowledgements

The author wishes to acknowledge the financial assistance of the NAWA Chair Professor Grant (PPN/PRO/2020/1/00003/DEC/1) from the Polish Academic Exchange Council and NCN grant (ZARZADZENIE NCN 94/2020) from the Polish National Science Council.

ORCID

Emilian Kavalski [iD] https://orcid.org/0000-0002-0364-959I

Bibliography

Adler, E. (2005) *Communitarian International Relations*. Cambridge: Cambridge University Press.

Allenby, B.R. and D. Sarewitz (2011) *The Techno-Human Condition*. Cambridge, MA: MIT Press.

Barkin, J.S. (2010). *Realist Constructivism: Rethinking International Relations Theory*. Cambridge: Cambridge University Press.

Beaumont, R. (1994) *War, chaos and history*. Westport, CT: Praeger.

Bernstein, S, R.N. Lebow, J.G. Stein, and S. Weber (2000) God Gave Physics the Easy Problems. *European Journal of International Relations* 6(1): 43–76.

Booth, K. (2005) *Critical security studies and world politics*. Boulder, CO: Lynne Rienner.

Bouris, E. (2007) *Complex political victims*. Bloomfield, CT: Kumarian Press.

Chernus, I. (1993) Order and Disorder in the Definition of Peace, *Peace and Change* 18(2): 32–51.

Cho, Y.C. and Kavalski, E. (2015) Governing Uncertainty in Turbulent Times. *Comparative Sociology* 14(3), 429–444.

Cioffi-Revilla, Claudio (1998) *Politics and uncertainty*. Cambridge: Cambridge University Press.

Clark, R.P. (2000) *Global Life Systems*. Lanham, MD: Rowman & Littlefield.

Clemens, W. (2013) *Complexity Science and World Affairs*. Albany, NY: State University of New York Press.

Connolly, W. E. (1991) Identity/Difference. Ithaca, NY: Cornell University Press.

Connolly, W. E. (1993) Political theory and modernity. Ithaca, NY: Cornell University Press.

Connolly, W.E. (2011) Complexity and Relevance. *European Political Science* 10(2), 210–219.

Cudworth, E. and S. Hobden (2011) *Posthuman International Relationss*. London: Zed Books.

Cudworth, E., S. Hobden, and E. Kavalski (2018) *Post-Human Dialogues in International Relations*. London: Routledge.

Davidson-Hunt, I.J. and F. Berkes (2003) Nature and Society through the Lens of Resilience. In F. Berkes, J. Colding, and C. Folke (eds) *Navigating Social-Ecological Systems*. Cambridge: Cambridge University Press, 53–82.

Der Derian, J. (1992) *Antidiplomacy*. Oxford: Blackwell.

Dillon. M. (2007) Governing Terror: The State of Emergency of Biopolitical Emergence, *International Political Sociology* 1 (1): 7–28.

Earnest, D.C. (2015) *Massively Parallel Globalization*. Albany, NY: State University of New York Press

Flockhart, T. (2006) Complex socialisation, *European Journal of International Relations*, 12(1), 89–118.

Fowler, A. (2008). Complexity Thinking and Social Development. *The Broker*, 7 April.

Frerks, George, Jeroen Warner, and Bart Beijs (2011). The Politics of Vulnerability and Resilience. *Ambiente & Sociedade* 14(2), 105–122.

Gettell, R. G. (1922) Influences on World Politics. *Journal of International Relations*, 12(3), 320–330.

Geyer, R. (2003a) Beyond the third way, *British Journal of Politics and International Relations*, 5(2), 237–257.

Grosz, E. (2004) *The Nick of Time*. Durham: Duke University Press.

Gunitsky, S. (2013) Complexity and theories of change in international politics. *International Theory*, 5(1), 35–63.

Haas, E. (1975) Is there a hole in the whole? *International Organization*, 29(3), 827–876.

Harris, P.G. (2009a) *Climate Change and Foreign Policy*. London: Routledge.

Harrison, N.E. (2006) (ed.) *Complexity in World Politics: Concepts and methods of a new paradigm*. Albany, NY: SUNY Press.

Heath, A. E. (1919) International Politics and the Concept of World Sections. *International Journals of Ethics* 29 (2), 125–144.

Heinzen, B. (2004) 'Surviving uncertainty', *Development*, 47(4), 4–8.

Heng, Y-K. (2010) Ghosts in the Machine. *International Relations* 47(5): 535–556.

Hoffman, M. and J. Riley (2002) The science of political science, *New Political Science*, 24 (4), 303–320.

Johnson, R. (2002) Defending Ways of Life. *Theory, Culture and Society* 19 (4), 23–47.

Kavalski, E. (2007) The Fifth Debate and the Emergence of Complex International Relations Theory. *Cambridge Review of International Affairs* 20(3), 435–54.

Kavalski, E. (2008) The Complexity of Global Security Governance. *Global Society* 22(4), 423–43.

Kavalski, E. (2009) Timescapes of Security: Clocks, Clouds, and the Complexity of Security Governance. *World Futures* 65 (7), 537–51.

Kavalski, E. (2011) From Cold War to Global Warming: Observing Complexity in Global Life. *Political Science Review* 9 (1), 1–12.

Kavalski, E. (2012a) Waking IR up from its 'Deep Newtonian Slumber,' *Millennium* 41(1), 137–150.

Kavalski, E. (2012b). Acting Politically in Global Life: Security and Its Logic of Resilience. In D. Walton and M. Frazier (eds). *Contending Views on International Security* (87–102). New York: Nova Science.

Kavalski, E. (2016) *World Politics at the Edge of Chaos: Reflections on Complexity and Global Life*. Albany, NY: State University of New York Press.

Kavalski, E. (2018). *The Guanxi of Relational International Theory*. London: Routledge.

Kavalski, E. (2020a). Inside/Outside and Around: Complexity and the Relational Ethics of Global Life. *Global Society*, 34 (4), 467–486.

Kavalski, E. (2020b). Complexity Thinking and the Relational Ethics of Global Life. Schippers, B. (ed.). *Routledge Handbook on Rethinking Ethics in International Relations* (11–24). Routledge.

Kavalski, E. (2021). Quo Vadis Cooperation in the Era of Covid-19. *World Affairs*, 184 (1), 2–24.

Kavalski, E. (2022). The Memory Bear. *New Perspectives*, 30 (2), 3–18.

Keohane, R. (2002) *Power and Governance in a Partially Globalized World*. London: Routledge.

Kissane, D. (2011) *Beyond Anarchy*. Stuttgart: Ibidem.

Korosteleva, E (2018) 'Paradigmatic or critical? Resilience as a new turn in EU governance for the neighbourhood', *Journal of International Relations and Development*, 23, 682–700.

Korosteleva, E and T Flockhart (2020) 'Resilience in EU and international institutions: Redefining local ownership in a new global governance agenda', *Contemporary Security Policy*, 41:2, 153–175.

Korosteleva, E and I Petrova (2021) 'From 'the global' to 'the local': The future of 'cooperative orders' in Central Eurasia in times of complexity'. *International Politics*, 58 (3), 421–44.

Kurki, M. (2008) *Causation in International Relations*. Cambridge: Cambridge University Press.

Kurki, M. (2020) *International relations and relational universe*. Oxford: Oxford University Press.

Mackinder, H. J. (1904) The Geographical Pivot of History. *The Geographical Journal* 23 (4), 421–437.

Manson, S. M. (2001). Simplifying Complexity, *Geoforum*, 32(3), 405–414.

Morin, E. (2004). The Ethics of Complexity. In Bindé (ed.) The Future of Values (43–46). New York: Berghahn Books.

Morin, E. (2006). Realism and Utopia. *Diogenes*, 39(209), 135–144.

Morgenthau, H. (1973) *Politics among Nations: The Struggle for Power and Peace.* New York: Knopf.

Murphy, P. (2000). Symmetry, Contingency, Complexity. *Public Relations Review* 26 (4), 447–462.

Nye, J. (1993) *Understanding international conflict.* New York: HarperCollins.

Odysseos, L. (2007). *The subject of coexistence.* Minneapolis, MN: University of Minnesota Press.

Pan, C. amd Kavalski, E. (2018). Theorizing in and beyond International Relations. *International Relations of the Asia-Pacific* 18(3), 232–241.

Petrova I. and Korosteleva E. (2021). Societal fragilities and resilience: The emergence of peoplehood in Belarus. *Journal of Eurasian Studies* 12(2), 122–132.

Prins, G. (1995) Notes towards the Definition of Global Security. *American Behavioural Scientist*, 38 (6), 817–829.

Rihani, S. (2002) *Complex systems theory and development practice.* London: Zed Books.

Rosenau, J.N. (1960). The Birth of a Political Scientist. *American Behavioral Scientist* 3(1), 19–21.

Rosenau, J.N. (1970) Foreign policy as adaptive behavior: some preliminary notes for a theoretical model. *Comparative Politics* 2(3), 365–387.

Rosenau, J.N. (1980). *The Scientific Study of Foreign Policy.* London: Frances Pinter.

Rosenau, J.N. (1988). Patterned Chaos in Global Life. *International Political Science Review* 9(4), 335–340.

Rosenau, J.N. (1990). *Turbulence in world politics.* Princeton, NJ: Princeton University Press.

Rosenau, J.N. (2001) Stability, Stasis and Change: A Fragmenting World. In R.L. Kugler and E.L. Frost (eds), *The Global Century* (127–153). Washington, DC: National Defence University Press.

Rosenau, J. N. (2003). *Distant Proximities: Dynamics Beyond Globalization.* Princeton, NJ: Princeton University Press.

Ruggie, J.G. (1975) Complexity, planning, and public order. In Todd LaPorte (ed) *Organised social complexity: challenge to politics and policy* (119–150). Princeton, NJ: Princeton University Press.

Wæver, O. 1997 *Concepts of Security.* Copenhagen: Copenhagen University Press.

Walker, J. and M. Cooper (2011). Genealogies of Resilience. *Security Dialogue* 14 (2), 143–160.

Waltz, K.N. (2002). The Continuity of International Politics. In Ken Booth and Tim Dunne (eds), *Worlds in Collision.* New York: Palgrave Macmillan.

Wendt, A. (1999). *Social Theory of International Politics.* Cambridge: Cambridge University Press.

Weber, C and Lacy, M (2010) Securing by design. *Review of International Studies*, 37(3), 1021–1043.

Whiteside, K.H. (1998). Systems Theory. *Policy Studies Journal*, 26(4), 636–656.

Zolkos, M. & Kavalski, E. (2016). The recognition of nature in international relations. In P. Hayden and K. Schick (eds), *Recognition of Global Politics* (139–156). Manchester: Manchester University Press.

'Imitated' or genuine? The value of resilience in Sufi-hamsoya

Nargis Nurulla-Khojaeva (iD)

Abstract *In seeking a framework for the study of resilience, this research turns to Central Eurasia, and more specifically, to the cultures of Central Asia with Sufis traditions. The chapter is premised on an imaginary dialogue of neighbours or 'hamsoya' ('sharing your neighbour's shadow' in Persian-Tajik). Following the hamsoya dialogue, we can trace the intergenerational culture of traditions without the pressure of delimited modernity and link this genetic intuitive knowledge to the future visions of the 'good life'. This is where the assumptions of genuine resilience emerge based on remembering (the verb from the epigraph; Quran 25:50). One of the possible ways of staying resilient is through remembering of who we are to each other which lies in the art of listening. Remembering recognises the human connections as central to life as the whole. It also helps communities in hamsoya survive and transform in the face of adversity.*

> *And We have distributed it amongst them so that they may remember, but most people refuse to be anything but rejecters Quran 25,50*

Introduction

In seeking a framework for the study of resilience, this chapter turns to Central Eurasia, and more specifically, to the cultures of Central Asia with Sufi traditions. While the question of resilience has been considered by many throughout the human history, the contemporary age, shaped by the experience of the global pandemics and other crises, has amplified and reminded us of the importance of this search. Notably, the COVID-19 pandemic pushed all of us indoors and posited many complicated questions for researchers: for example, are we able to overcome the barriers of exclusion and isolation, to 'return' to our resilience and *remember* it for what it is—as self-organisation and community of relations allowing humans to survive *together* in the face of adversity?[1] Why is it so difficult to 'return' and *remember* what it was like before modernity? If we are able to remember, can we rediscover the true meaning of resilience then?

These multi-layered questions, and their very formulation, entail complexity and uncertainty. In my academic intuition-driven thinking, this appears in a stuck contrast to rational thinking, or the Cartesian picture of the world in its simplistic binary of the subject–object relations, which *irrationality* has been exposed by the COVID-19 pandemic. Even the best-thought actions are sure to be constrained by the impossibility of planning or exerting control. The Sufi philosophical thought captures this impossibility in a familiar way: there is an abyss between the man and the Other; between the man and the World. In what follows, we will explore the beginnings of these relations, from a tradition of Sufism, and why it is important to remember resilience as *hamsoya* in the world of today.

The global COVID-19 crisis demonstrated to us time and time again, that linear thinking imagining the world(s) as a closed system, no longer works. In the search of an alternative, this chapter suggests to return to *remembering* the era of 'engagement' when people felt and shared resilience as a communal property. Prior to the pandemic this linear thinking had remodelled International Relations (IR) making us see and understand the world in a very Cartesian way, rather than in its multiciplicty. For the majority, IR still remains a study of borders and influences of one state on another (vaccine nationalism is a clear example of this in our age). Because of this we lost a sense of 'relations' themselves, as IR is not about developing a 'sense of cultural unity' (Zargar 2017) between the neighbouring states. On top of this, there is a clear blind-folded disposition of many westernised academics, who also tend to view the world in a linear way of 'the West over the Rest'. This modernised and secularised worldview ignores the uncertain, the complex and the God-like nature of this world with everything and everyone being inter-related and connected to each other. With this singular-minded approach, the state as an institution penetrates the domains previously run by non-state agencies (education, health, environmental protection) which were always oriented towards a greater vision of unity and connectivity—'the unity of the ultimately Real' (Chittick, 2011, 88). What if the COVID-19 pandemic has challenged this linear thinking and pushed us to 'return' and 'remember' the era and the art of 'engagement' and togetherness?

The word over the deed

One way to understand resilience, at least in the philosophical traditions of Central Asia, is by turning to the art of listening, or what Sufis call *sama*. There is a double mysticism in *sama*: one can listen with both ears and the heart. The heart has an 'ear', and it reflects a voice, sometimes only the intonation of the Quran's recital, or even the movement, gesture, rhymes, and the heard word. Indeed, everything can contribute to the spiritual uplift of a Sufi. The primacy of listening stands in direct opposition to the idea that 'in the beginning was the deed', as worded by Goethe's Faust and reflected in the book of Bernard William under the same name (2005). These 'deeds' are replacements for thoughts and their primacy is associated with the current rigid system of our knowledge and education. This runs counter to the *genuine* capacity of resilience, based on emotions and relations, and driven by idea that 'truth is transcendent' (Bashier 2004), and it 'never assumes a single form' (Wood 2017). However, saying this does not mean that we are declaring that 'there is no such thing as a uniform temporal medium' (Anderson, 1974, 10). Instead, it may simply be about 'returning to the root', keeping in mind that 'Sufism and Philosophy are Neighbours and Visit Each Other' (Rosenthal 1988). A Sufi and a philosopher, as Neighbours, can explain the meaning of modernity, without getting mired in the limitations of the state, nationalistic outbursts, religious wars and juridical fault-finding, striving instead

towards the pursuit of knowledge, as declared in the Quran. After all, these two—
a Sufi and a philosopher—have trust in Others.

Trust is perhaps the most essential component in maintaining resilience
among them. Paradoxically, Western scholars have only recently begun to probe
what trust is, and how it affects the state and international relations (Haukkala
et al. 2018). However, in such an alignment, the clear interest in state structures
institutions and their nationalistic tendencies negates the very meaning of trust
among 'neighbours', thus challenging any form of resilience and relations. It
is here that I wish to ask the reader to ponder over the challenges of defining
resilience within this conventional border-thinking matrix. The intuitive process hidden in resilience cannot be unfolded or framed, because every definition
would reveal the multitude of meanings and also limitations that cannot be simply
fitted into a neat narrative, as is argued by Chandler in this volume. Here, in this
moment of reflection, I turn to a Bukhori (a resident of Bukhara city), whom
I met during one of the conferences in 2019. More specifically, while visiting the
Sufi shrines around the city, a stranger kindly quoted an old proverb to me: 'As
plums are similar to each other by colour, so neighbours take good thoughts from
each other.'[2]

The 'good thoughts' notion and its primacy among neighbours remind us
about one of the ancient and long-standing traditions of Central Asia, namely
the belief in the magical power of 'Good Thoughts, Good Words, Good Deeds'[3]
from the Zoroastrian text of the *Avesta* (9th century BC). The primacy of thoughts
and words over deeds is also reflected in the Logos of Ancient Greeks, and in
the Gospel of John: 'In the beginning was the word', because each one's 'good
thought' stands with the 'good thoughts' of Others. Such commitment cannot be
defined in terms of the rigid propositions of the Faust-William type, favoured in
IR with the preset national borders, and social, political and religious divisions.
'Good thoughts' rely on the trust of people, who are neighbours and necessitate such 'thoughts' to affirm the much sought-after resilience. This feeling is
also supported by the etymology of the word neighbour: according to my Uzbek
colleague, the word for 'neighbour' in his language is *qo'shni* (constructed from
the word *qo'sh*, meaning 'together'), which has a similar word in Azerbaijani, *qon u*
(where *qon* also means 'together'). We deal with things in terms of the names
that we give them, which if anything, demonstrates the primacy of the word and
the art of listening over the deed. My Bukhari stranger used the word *hamsoya*
(in Persian-Tajik *ham* is together, *soya* means shadow) which means those who are
'united in the shadow' of the neighbours, trusting and listening to each other,
before putting their words to practice.

What if I become the reader's *hamsoya* and we strive to share the 'soya' ('shadow')
during this conversation? Most of us are in need of conversation—a conversation with neighbours, without the modern feeling of individual 'subjectivity' and
competitiveness, with a chance to humanise our communication. Our current
sedentary status partially inflicted by COVID-19 will probably enable us to call
such a neighbourly talk as a 'deconstruction of presence', as Derrida (2016) once
said. This way we could develop a discussion through the prism of poetry and

Islamic philosophy, when a 'community without a community' is created, based on Rumi's principle of listening with the heart. This is where 'I am You' and we are *hamsoya*—we are those who are trying, together, to find inspiration from those who are trusted, who may be nearby, or beyond time and geography. Such an imagined setting allows accepting Sufi-*hamsoya* into our conversations, together with the *hamsoya of the Other*, for whom history and modern borders are non-existent and irrelevant.

In the Persian-Tajik language, along with the word *hamsoya*, the prefix 'ham' appears in many other words, notably *hamroh* (united by road), *hamdili* (united by heart or solidarity), *hamrang* (united by colour), and *hamdard* (united by pain). As the underlying etymology of words related to *hamsoya*, the shared prefix 'ham', can be taken as marking 'togetherness', which is particularly well-noted in poetics. One can grasp how long and meticulously the 'I' was separated from the others to feel *hamdili*. Through 'ham', we might demonstrate our resilience as a way of thinking and listening to others (see Korosteleva and Petrova, Introduction) which is historically 'always more' than what current policies and practices of resilience are providing (Bargues-Pedreny 2020, 265). 'More' assumes an important role in stimulating community of relations or 'neighbourhood' as we have come to define it here.

At the first glance, this approach may strike you, my reader and my new *hamsoya*, as another platitude that the ambitious pluralists bandy about. Yet, this idea of relationality and togetherness can, and should serve as the foundation for reshaping our thinking today, as part of returning and remembering that resilience is a process of horizontal relationship-building and keeping, including through the imagined conversations in *hamsoya*.

The Sufi premise

At the heart of poetics in Sufi traditions lies a special, respectful attitude to the word, to its use, a kind of cult of the word, through which 'an elegant compromise has been reached between the constraining requirements of a finite religion and the irresistible human urge for the infinite exercise of the imagination' (Carter 1998, 240). Probably, for this reason, the majority of poets of the Islamic languages of Eurasia, including Persian, Arabic, and Turkic, were Sufis. In their poetry, the pathos of a hint, a play with circumbendibus and allegories, the ambiguity of half-tones and chiaroscuro have long been dominant. Accurately, one should start with the name of Jalal ad–Din Rumi (1207–1273). According to the latest research, he was born somewhere between modern-day Afghanistan and Tajikistan, travelled extensively within and outside the region before settling in Konya, in present day Turkey. However, in our case, modern-day geography should be set aside, as we have *Mavlono*[4] Jalal ad–Din Rumi, whose existence extends far beyond the borderlines and whose *Mathnawi* deals with *usul usul usul al-din* (from Arabic, 'the roots of the roots of the roots of the religion') where religion is something more than just 'religion'. Rumi was the one to shrewdly pointed out that 'outwardly we are ruled by these stars, but our inward nature has become the ruler of the skies' (Rumi 2012, 54). Such circulation between 'we' and our 'inward nature' gives us an understanding that the object and the subject of knowledge are indiscernible and inseparable: it is about an genuine unity of the individual, on micro and macro level; that is to say, it is about togetherness, the *ham*. This basis enables me to extend the invitation to you, the reader, once again, to become a member of *the hamsoya* community, or 'Rumi community', which is a familiar way of thinking, listening and sharing your thoughts with a neighbour in Central Asia. This community can explain the historical and political conditions of the current world and the seemingly separate worlds of the East and the West, not in a linear way, but importantly, 'rather in terms of neither/nor or both/and' (Bashier 2004, 61). This approach enables exposing the limitations of the existing systems and deconstructing the prejudices therein. As such, this conversational approach allows presenting Islam not as a frightening network of *mujahideen*, but rather as a 'community of discourse', including 'the verbal, the formal, as well as the informal, the gestural or ritual, as well the conceptual' exchange (Wuthnow 1989, 220). Even more, this communitarian platform free of any emphasis on hierarchy and binary opposition's works as an antidote to our attempts to recreate strict borders among *hamsoya*. The platform might remind one of Glissant's concept of Relation, as 'it is not a language of essence; it is a language of the Related' (Glissant 1989, 241). Long before the latter, Rumi had voiced this idea: 'I look not at the tongue and the speech; I look at the spirit and the state (of feeling)'.[5] Perhaps, as my *hamsoya*, the reader might agree that the Rumi-Glissant picture of 'relations' explored in this volume by Chandler and others, organically delivers a more complex world to encourage our relational being in it.

The storyline of this chapter follows a set of issues ranging from the unnecessary rationalization of resilience by modern IR scholars and the institutions of higher education in general, to the questions of what *genuine resilience* is, when it is premised on the integration of 'knowledge', emotions, *hamsoya* and their symbols, this way showing the interwovenenss of these concepts to better understand the world. These and other shadowing questions signify a present-day confusion in our understanding resilience, and encourage us, as *hamsoya*, to feel the in-betweenness of resilience and the importance of remembering what it means to be together. The thorniness of this incredibly important and mystical vision of our world is simply intensified by COVID-19, so much so that it probably fastened our understanding of resilience as an art of listening to the world and the self.

'Imitated' resilience and adab

For some of us, the process of understanding the many meanings of resilience relates to traditional research with a certain normative content. We know that norms and rules continue to regulate our living. Everything that is interpreted as 'irrational' or an outlier of the accepted norms is brushed aside (e.g. to the fringes of Islam, ethics, etc.). Most of us probably know that our working syllabus is designed as a 'reflection [that] can only roam in its own specific playing field (*maydan*)'. Likely, 'one half of the term is artificially privileged over the other for ultimately ideological purposes' (Almond, 2004b: 24). The caution comes from Al-Andalus, another margin of Eurasia (distant from Central Asia), the place where the great Sufi philosopher Ibn Arabi (1165–1240) was born. He was called by other Sufis 'animator of the religion' and *al-Shaykh al-Akbar*—the Greatest master. In our *hamsoya* conversation, he is able more than other philosophers to prevent me and you, my reader, to become addicted to numerous divisions and blame games. Ibn Arabi can animate such a process because '...The intellect restricts and seeks to define the truth within a particular qualification, while in fact, the Reality does not admit of such a limitation' (Arabi 1980, 150).

However, one can adapt to living with limitations. The modern matrix of university knowledge is constructed as a process of learning the 'subject' here versus the 'object' out there. When I studied at the university, it was named after Lenin (the name was kept till 1997), and of course, it had a complex Soviet legacy. I received my diploma believing that nationality is inevitably associated with the ancient history of the modern Tajik and Tajik Soviet Socialist Republic. My socio-political interpretation of Tajikistan's history however reflected the state's interests which form the backbone of contemporary ideology. With my regional *hamsoya* (the Soviet generation), I was involved in composing a unique Union of fifteen national units, which was an innovative mechanism of new opportunities that unfolded the trajectory of the region's development for a long time. With the dissolution of the Soviet Union, our *collective hamsoya* was forced to return to their national homes, without remembering the common space of thought and trust built.

My Sufi-*hamsoya* suddenly felt sorry for me; he realised that education in modernity has a different arrangement. In pre-modern times teaching, in general, including the network of intellectual activity, was neither on the margin nor the jurisdiction of any political power (Hallaq 2018). In those days, intellectuals believed that intellect could be 'partial' (what Rumi called '*aql-i juzwi*') and 'Universal' ('*aql-i kulli*'—'Universal Intellect'). The first type of knowledge is not on rationalisation only; it is a synthesis of the rational, poetical, and spiritual guidance of a Sufi student (*murid*) from a Sufi master (*pir*), from his ethical model. But it is, at the very least, blinded by self-interest, insists Sufi-*hamsoya*. The second—intuitionally Universal (*aql-i kulli*): it is when an individual is free to define capacity, to intuitionally conceive the truth within Self, without accepting God's instruction on what to think because 'that would be to make imitation and hearsay incumbent in intellectual matters' (Chittick, 2009:6). Such context and dialectical traditions of a constant process of learning allowed a Sufi master and his student go through the stages of actualisation. They can *remember* a compromise by crossing the antinomies between Nature and Self (with no attempt to pretend to satisfy aesthetic and ethical needs). This is a complex way of intuitionally reaching *aql-i kulli*: a Universal Intellect and *genuine resilience*.

Nonetheless, I am not Sufi; I am a researcher who cannot recapitulate resilience in modern times. Do you think it is the right time to ask myself about my incompetence to comprehend resilience as *hamsoya* or declare its ending because of the 'uncertainty' of Science (Chandler, 2020)? What if today's context illustrates that my mistake (including many other experts) is to handle social and moral values associated with resilience as scarce commodities? Or, what if it simply indicates the shortage of compromise and intuition? I know... and agree on a priory with my *hamsoya;* these questions sound strange, but do you agree that the picture is too complex to reconcile with our 'objectively' established academic knowledge? Do you think Rumi is right to say that my *particular intellect* has disgraced [Universal] Intellect (Persian-Tajik: *Aqli juzvi aqlro badnom kard/ Komi dun'e mardro nokom kard*)?

In the rational vision of resilience, through the liberal theory of good life, we think that 'risk cannot be 'reduced' as all actions have interactive outcomes that need to be assessed' (Chandler 2018, 45). Do I need to buy into this mechanical liberalism with Fukuyama's 'end of history' in any case? Or is it better if I try to comprehend and value my *hamsoya*? Given this confusion, would it be right to refer to classical Sufi poets, who developed and expanded the conception of a good life differently? For instance, Saadi Shirazi (1210–1291), whose poetical heritage is well-known to the Persianate Eurasian world, said: 'The sons of Adam are limbs of each other, Having been created of one essence' [In Persian–Tajik: *Ba'ni odam a'zoi yakdigarand/ Ki dar ofarinish zi yak gavharand ....*]. The poet reminds us that *Ba'ni odam* (Persian-Tajik: *Human*) is not a 'citizen of the world', and the poet does not consider a different version of the good life for some *hamsoya* only. In Saadi's case, the concept of a good life is good when me-you share 'shadow' as a constant processing of humanness, good practices like in old neighbouring communities of ancient Eurasian cities, such as Bukhara, Samarkand etc. It is evident that 'at the heart of any global or local arrangement, and even within the individual person, is a reflective notion of the "good life" underpinned by "normative beliefs and century-long traditions that shaped the ways of life for local communities and individual human beings".' (Korosteleva & Flockhart 2020, 163) It means that 'the *character* of the order was more important than its *form*: it should neither be repressive and static' (Moosa 2018).

The *form* of a good life, its meaning and weight, is associated in different parts of the world, including Central Asia with *adab* (loosely might be translated from Arabic as 'upbringing' or 'educating'). 'This cultivation, far from being an externally imposed system of 'training the subject,' was designed to operate internally, through deliberate and self-conscious choices of conduct and belief that the autonomous subject exercises upon his or her body and soul.' (Hallaq 2018, 75). Like our favourite resilience, the word *adab* is complex to define. It resists a solid definition: the first figurative meaning of *adab* is a 'graceful speech', and the second is a 'good speeches and good deeds'. *Adab* requires good practices via poetry, aphorisms, proverbs, symbolic images, illustrations, knowledge of history, and the correct and attractive use of language.

The concept's semantics and meaning are rich thanks to the adoption of numerous multi-ethnic historical and cultural passages of ethical behaviour and cultural exquisiteness in the Medieval Islamic socio-cultural environment. It means that *adab,* is linked to the history of ideas; not bounded by the boundaries of culture or politics, and having some universal shadow. A famous Tajikistani composer, Ziyodullo Shahidi (1914–1985), had in mind *adab* while explaining a centuries-long tradition of ditches and open-air water reservoirs. Based on *adab,* a respectful approach toward water use, for example, allowed each community (also known in Samarkand as *guzar*) within the city to receive clean running water. In return, member of each *guzar* was responsible for the removal of excess water from the city, and at the same time, among people in all communities, a reverent attitude to water remained; respect for water itself and water in the canals, one's water and that of *hamsoya*. The complex part of the explanation of Ziyodullo Shahidi was the nuance that he elucidated *adab* as a beneficial object. Just later, I understand that in Islamic terms, a benefit must be designated by the ultimate principal issues, closer to *remember* through essence. It hints that disturbances and breaches of *adab* no doubt existed, but such incidents never amounted to be considered normal. At the end of our brief talk on *adab*, my Sufi-*hamsoya* whispers, if you want to get the meaning of genuine resilience, it would be right to go back to poetical fluidity based on *adab* in our university environments.

The 'seven birds' on the Coat of Arms of hamsoya
Islam is perhaps one of the few cultures that have chosen diverse ornaments (poetical, artistic, imaginative) as the most important means of identification, endowing it with the extensiveness of divine semantic images and associations. The trend might be illustrated by the words of the representative of Tartu's School of Philosophy, Yuri Lotman (1922–1993): 'to be active, consciousness needs consciousness, the text [needs] text, and culture [needs] culture' (Lotman 1981). It would be reasonable to add that *hamsoya* needs *hamosoya* as well.

During the medieval time, the importance of social norms and informal social design of *hamsoya* community was based on a productive relational balance of 'text and context', or a holistic sacred concept of world harmony. Such composition in Islamic art was known as *bandi-rumi* (Persian-Tajik: the *band* is a knot and -*rumi* is Eastern Roman Empire or Byzantium), that is, Byzantium's ligature. It directs us to a theory of multiple times, 'eternal prototypes' of the genuine resilience of human beings.

From this standpoint, we can introduce an important manifestation of *hamsoya*—the symbol. A powerful meaning can be found in the image of a bird in Eurasian space; it is associated with divinity, immortality, power, spiritual victory, and hence—royalty, God-given power over people in Eurasia. Sufis devoted many poetical treatises on and around Birds, giving the role of the spiritual locus to Truth-God by Sufis. For them, the transformative nature of the character is such that it provides access to a realm beyond rational thinking where rational exclusions and boundaries cease to define reality. One of the birds *Huma* (sometimes has a different name: *Simurg or Anka*) has the highest prestige. The legend assured a lifetime of happiness to the person who happened to be under the shadow of *Huma*: 'the bird foreshadows the happy life' [in Pers.: *murg-i humay' un-bol*].

Enigmatically, a bird's soul from Sufis poems and miniatures managed to cross borders of time and cities. It landed in modern Uzbekistan, Kirgizstan, in their Coat of Arms.[6] Needless to prove that their *hamsoya*, Tajikistan, has a Coat of Arms linked to the ancient times. The frame of the blazon of Tajikistan is crowned with the tiara with seven stars placed above it in a semicircle. The anagogic meaning is hidden in the 'seven'.[7] The seven-gods pantheon was an ancient Aryan template associated with 'seven planets ... which are in interaction with 'mothers' (four elements) [wind, dust, water and fire] perform a major role both in cosmogony and in the fate of humans (microcosm)' (Abaev 1962, 447). The case of 'seven birds' was described by Yuri Lotman. He wrote: 'An immutable set of symbols passing diachronically through a culture, [it] assumes to a significant degree the function of unifying that culture' (2019, 135).

In Central Asia, we know that our modern visible perceptions, such as nationalism, progress, atheism, socialism, individualism and plenty of other -isms, arrived at our place after 'Seven Birds'. Of course, recognition and sometimes admiration -isms are our fault because 'the apparent discontinuity between individuals, the relation of their respective centres of consciousness, is only the mark of their unique Essence which 'vertically' transcends the 'horizontal' plane of their common nature...' (Burckhardt 2008, 26). We do not remember how our 'birds' used to connect members of intellectual communities' network of cosmopolitan cities, like Samarkand, Bukhara, Khujand (Central Asia), Ganja, Iravan (or Yerevan, in the Caucasus), Konya, Istanbul (Turkey), Delhi, Calcutta (India), Kashgar, Hotan (China) ... The resilience of those city intellectuals used to promote patterns of acculturation to interweaving religious-literary-commercial-spiritual-manufacturing traditions and the social 'togetherness' of hamsoya. It is about rememberring and mixing very different histories and geography. For instance, the revolution of 1917 in Bukhara (Aini 1987) was opposed to the modern westernised statehood institutions. It had fewer accents on political divisions and more on community commitments. But paradoxically, Bukharan case might guide us back to a Marxist philosophy (closer than some researchers believe). It is known that after the Paris Commune (1871), Karl Marx changed his vision of the phenomenon of the state; he stopped focusing on the issue of replacing one state with another (the revolutionary dictatorship of the proletariat), clearly presenting the state as a 'despotic' institution. According to the thinker, the commune broke the power of the modern state and its despotism and thereby tried to 'reabsorption of the State power by society as its living forces instead of as forces controlling and subduing it' (Marx, 1985). Marx's interpretation of communal-hamsoya meaning can cause, to put it mildly, confusion. Still, it can provide 'a better sense of what kind of multi-level governance is needed' (Korosteleva and Petrova 2022, 3, and Flockhart in this volume).

Comparability and correlations in different scientific and geographical areas were discussed at the international conference hosted in Dushanbe on 'Tajikistan and Cultural Diplomacy in Central Asia and Eurasia' in September 2019. Members of different University-partners and the Tajikistan National University team came together to highlight the interconnectedness of the Central Asian region and define its holistic design. It was the first time my university had welcomed a talk on Cultural Diplomacy and the meanings of connections with hamsoya within the region, around the Eurasian region and the world. We acknowledged that, on the one hand, intercultural diplomacy is about the modern state culture and ideologies-*isms* which '...can explain everything and every occurrence by deducing it from a single premise' (Arendt 1953). On the other hand, many confreres underlined that the centrality of Central Asia and its internal interculturality might also help broaden interpretations of -*isms*. My *hamsoya* perhaps agreed that 'seven birds' can be an 'object' and a 'subject', and a symbol of this region's united intercultural traditions. It follows that there can be no 'bird' without a network of intellectuals-*hamsoya* and no *hamsoya* (who united by shadow) without a 'bird', or invisible dimensions of a dream of universality or universal resilience among us.

Such an arrangement makes me think that Charles Kupchan's idea that there is 'No One's World' is partly true (Kupchan 2012). But it is only partly. What if genuine resilience is about No-One and One at the same time: no West, no East, and they are synchronically together? It is comparable to what Derrida called deconstruction: 'tension between memory, fidelity, the preservation of something that has been given to us, and, at the same time, heterogeneity, something new, and a break' (2017, 29). Though Derrida's 'break' reminds us not of a 'new' resilience but facilitates to *remember* a genuine one. It is presented on a symbolic-poetical stand because everything is interconnected around me and my *hamsoya*. We have a preternatural shadow, which infiltrates with bird-soul.

A dream of resilience, and self-hamsoya

Suddenly my Sufi-*hamsoya* continued his test on me. He said that if I were *murid* (student in a Sufi league), veridical resilience is hardly reachable for me or someone who was educated in the modern university. A similar diagnosis as 'cultural schizophrenia' was framed by Daryoush Shayegan (Shayegan 1997). Perhaps there is truth to this. My education is based on achievements of definitions, and I cannot fit spirituality in my academic structure; it is outside of it. All my professional training was solidified by history, ethics, theology, philosophy, and law courses. Those Western disciplines circulate their concepts with a non-Islamic pedigree; they are based on a method that entails a dangerous generalisation of religion, particularly Islam. Ibn Arabi adds drops of pessimism, presenting my knowledge as *iqal* (Arabic, *chains*). Of course, for me, it would be easier to say that the system itself is colonial since it reflects a constant *divide et impera* principle; demonstrates the spirit of dyadic logic when 'I am a Subject and Nature is the Object'. I can blame Kantian reasonings (or his paradise arguments), 'Yes' to the rational, and 'No' to ir-rational. However, here I have a warning from Ibn Arabi, with whom we converse to avoid the object-subject divisions. We all have enough regulations, borders, and priorities on our shrinking planet. But... what if the pandemic pushes us to *remember* to understand that the issue is in an inability of Self to think from the standpoint of *hamsoya*. It is a reappearance of Ibn Arabi's answer, his way of connecting 'Yes and No'. This exact was to respond to the question of whether an understanding of the 'divine radiance' is achievable rationally. The response might aid us in thinking about what we should do when everything depends on understanding what we are, as *hamsoya* or people who share the 'shadow' of human togetherness. If we *remember*, in that case, the epiphany of decolonisation or another, a universal vision of Self comes by itself. Such an approach is very close to Ibn Sina's philosophy. I recall how often Ziyodullo Shahidi referred to this out-standing polymath of the X–XI centuries from Bukhara. Ibn Sina is the most famous *Bukhori* (resident of Bukhara-city), whom experts called the 'philosopher of contingent being'. He modified and enlarged the 'First Teacher' (Aristotle's title) accession by declaring that 'these are the assumptions which are warranted neither by reason alone nor by senses alone, but which can be known by the two working together' (Ibn Sina 1983, 25). The modifications to Aristotle's ideas were presented mainly in *Kitab al-Shifa* (Book of Healing), the second encyclopedic book of Avicenna. The title is proper because the first one was devoted to healing the body (The Canon of Medicine), and *Shifa* is about the Soul. In that Book, Ibn Sina visualised the first virtual experiment of humanity—a 'Flying man' (some-times called 'Soaring' in Arabic: *shu'urbi-al-dhat*). We could do such an experiment together if we joined him, as *hamsoya* (he was Bukhari, after all). Can you recall the 'Vitruvian man' of Leonardo Da Vinci (1452–1519)? It would be Avicennian's 'Flying man' or Human, who does not have a possible form of intelligent percep-tion, no previous experience; it is an extreme case of perceptual isolation (or sen-sory deprivation). If Aristotle believed that all knowledge arises from sensations, then the 'Flying' has a stable awareness of his consciousness; it instead occurs together with himself, with his recognition of himself: through himself and from himself. It resembles Socrates' stance: '*I know that I know nothing.*' Here comes the recognition that self-awareness occurs not a Cartesian way. But here we have a traditional round of questions: when I see myself in the mirror, is it a reflection of physical form only? But what about my self-awareness? Is it there? Does it reflect features of 'Flying'? Does my Self talk to myself? Can I go away from myself? Reflections might assist in being alert that awareness is grounded on a priory imprint and operates as an epistemic base in Self-awareness.

We have a chance to present proofs of such a complex Self-awareness design via references first to a custom of Ibn Sina's land, his Bukhara and then to Goethe, his 'West-East Divan'. In both cases are possible to track a particular understanding of the epistemology of the unity of being, where, balancing between experience and intuition, Self and your *hamsoya* lays the foundations of a dream of universality, or what we might call a *genuine resilience*. Bukharan case: upon the completion of a wedding ceremony in this ancient city, a groom and a bride are shown to each other as a united couple through the mirror for the first time as a married couple. Later, the eldest woman in the family, who holds the most honourable status in the family hierarchy, greets a newborn baby and puts a small mirror under the baby's cradle (Persian-Tajik *gavhora*). Another example of a "mirror" comes from Germany. 'Almost two centuries before, Goethe (1749–1832), in his 'West-East Divan', offered an idea of 'poetical twins', describing himself and Persian poet Hafiz in the dialogue of 'seen' and 'unseen' poets' (Shahidi 2016). It is an intui-tive picturing of a virtual mentor (Arabic *khidr*), master (*sheikh, or pir*), friend and beloved in the heart.

In all mentioned cases, partakers (Self-awareness with 'Flying man', sharers of mirrors traditions in Bukhara, Goethe with Hafiz) are presented with a constant crossing status. I think for each of us: I and my *hamsoya*, closer than

Self-awareness, have nothing. We know this is a prevalent feeling for all people. Ibn Arabi complemented the image, saying: 'If he says: 'I saw my form I did not see my form,' he will be neither a truth-teller nor a liar' (Bashier 2004 18). In some ways, we have his classical 'Yes and No', two in one, or Self-*hamsoya* here.

It is probably time to comprehend that Sufi and Philosopher are *hamsoya*. This couple told us that preserving humanity through contemplation and love is our shared responsibility. If expected, it would be fitting to *remember* that the word *insan* (from Arabic) is known to many people of Eurasia. The word means 'human'. Ibn Miskawayh (932–1030) said, '*insan* from '*uns*, friendliness, sociability treating the idea of humanity as the derivative notion and sociability as the basic one, rather than vice versa, as the linguistic givens might lead us to expect' (Goodman 1999, 132). In the word, we have our treasured genuine resilience again: when we are together with others, and neither or nor against, that is human togetherness and what we might call *hamsoya*. The approach allows avoiding the central mark of modernity—a spirit of competitiveness, recognising diversification. The word *human,* in its turn, is, coming from the Latin language, rooted in 'earth, ground; it can be traced to the Proto-Indo-European root *dhghem* (eath) (Krell 2020, 173). I am sure for all *hamsoya* is biblical: 'Dust thou art and to dust thou must return' (Bible).

In Dushanbe (Tajikistan), it is poetically biblical since the circle of Human life is audibly presented by Omar Khayyam (1048–1131) with the question: *I went last night into a potter's shop, / A thousand pots did I see there, noisy and silent; / When suddenly one of the pots raised a cry, / "Where is the pot-maker, the pot-buer, the pot-seller?"'* (transl. by Cowell) [In Persian-Tajik: *Dar korgahi kuzagare raftam dush, Didam du hazor kuza guyevu homush. Nogo yiaki kuza bar–ovard khurush 'Ku kuzagaru, kuzaharu, kuzafurush.'*] However, today's Dushanbe in my University do not *remember* that we had a musical extension of Khayyam's question. It is Tolib Shahidi's ballet '*Rubai of Omar Khayam*'. Ballet thundered in the 80s last century throughout the vast Union of 15 republics. It might be breathtakingly for my *hamsoya* to see how the symbol of the delicacy of human existence, the pot breaks down again and again during ballet performance, the same fragile pot moulded by Divine Potter. Nevertheless, at my University, we remember cartesian philosophy: Descartes leads Self to the idea that 'my body' is just an illusion; through this prism, he invented 'consciousness'. Based on this, the savant initiated the dominance of physics, believing that this discipline (together with geometry) could provide us with knowledge of matter and its properties. This skillful avoidance of scholasticism becomes the main link in constructing European monistic experimental science. However, don't you think such a view leads us to simplify humanness and its nature and resilience?

Perhaps Ibn Arabi is right to infuse: 'Never assumes a single form' because 'increasing science is not increasing demystification' (Morton 2013). But what does the inability to think about Humanness means? Rumi answered that question: *Do you know why the mirror does not reflect? / Because the rust has not been removed from its face* (transl. R.A. Nicholson) [In Persian-Tajik: *Oina-t doni charo qammoz nest, / Z-on ki zangor az ruhash mumtoz nest*]. The mirror of a heart, the organ of 'mystic physiology', is a part of the ancient magical technique implemented by the Sufis into the sublime vision of *tawhid* (Arabic: *the oneness of God*) through Self. But what if Humans stop 'polishing heart' and stop thinking? How do learn to think by the very views of humanness?

One of the options is suggested by Ibrahim Moussa. He wrote that perceiving of Sufi-philosopher (he meant Al Ghazali) requires from researcher 'to be both an ethnographer and a native at once, or .... be both observer and witness at once. As an ethnographer and observer, one derives meaning in a cultural-linguistic frame; as a native and witness, one searches for meaning in an intratextual hermeneutical frame' (Moosa 2005). These interpretations make it possible to initiate a new discussion of *hamsoya* for possible ways to bridge the gap between East and West. The approach does not imply the boundaries of spheres: where there is independent thinking and separates beings; it does not determine preferences and obligations (state, nations, minorities).

Further thinking

As an influential figure in the post-colonial study, Bhabha said that such a person as me (who lives not in a born city, in my case, not in Dushanbe) lacks the primordial unity or fixity (Bhabha 1994). And yet it is illogical to insist in our post-pandemic time on the primacy of one theory, albeit as significant as a state within the national borders. At this point, should we agree that the *Human* and *Insan* could really be seen as a reflection in the mirror? Our review suggests 'sameness' also amounts to 'dissimilarity'. This is a constant crossing and the enrichment of what and who we are in our *hamsoya* to each other; we inevitably need to transfer what we see in ourselves to the outside world, but also to absorb everything important from it too. From here, the idea of a microcosm (body) and macrocosm (reality around) and their connection; because 'Truth is transcendent' (Bashier 2004) and our human togetherness has a third space, the space that does hold resilience. *Remember* is about acknowledging that third space, when each of us genuinely dreams to enter in and out. But if we decide to stay in a line with *rejecters* (Quran 25, 50), then unlikely we are ever to understand that the shared symbolic practices and languages, appreciation and love need to be restored because 'He gave us divinity/We gave Him humanity' (Ephraem 2015, 87).

Munira Shahidi from Tajikistan National University remembers these words of Ephraem the Syrian (d. 373). In her lectures, she tells her students that the son of the Virgin Mary is the most visible person in the Qur'an (mentioned 187 times). I know that the Professor usually adds a few lines from Hafiz. The outstanding poet, in his turn, testifies to the liberation of Christ himself from the borders which we created for him when the Son of the Virgin Mary 'danced' in the rhythms of Venus, the symbol of love (the early Christians prayed for this planet): '*No wonder if in the heavens the words of Hafez/ Brings Venus to singing and Christ to dancing sprees*' [Persian-Tajik: *Dar osmon na ajab gar ba guftai Hofiz,/ Surudi Zuhra ba raqs ovarad Masehoro*]. If we try to *remember* to listen not to politicians but poets, then we have the potential to learn to hear each other's *hamsoya* with an ear of the heart (*sama*). In the end, Rumi settled to all *hamsoya*:

You do not see the applause of the leaves / For this, it is necessary not the ear, but the ear of the heart. [Persian-Tajik: *Tu nabini barghoro kaf zadan, / Gushi dil boyad, na in gushi badan*].

Acknowledgements
The author wishes to thank the editors and the anonymous reviewers for their helpful comments on the earlier drafts of this paper, and the PI of the GCRF COMPASS project (ES/P010849/1), Professor Elena Korosteleva, Warwick University, for leading this exciting project.

Disclosure statement
No potential conflict of interest was reported by the authors.

Notes

1. See the introductory chapter by Korosteleva and Petrova in this volume.
2. In Persian-Tajik: [Olu zi olu girad rang, hamsoya zi hamsoya girad pand]
3. In Persian-Tajik: [Pindori Neck, Raftori Neck, Guftori Neck]
4. The term *Mavlono* (or in Arabic, Mawlana, meaning lord) is a title, mostly in Central Asia and the Middle East, preceding the name of reputable and respected Muslim spiritual leaders.
5. In Persian-Tajik: [Mo zabonro nangaremu qolro, mo ravonro bingaremu, holro]
6. www.heraldry-wiki.com/heraldrywiki/wiki/Uzbekistan
7. www.heraldry-wiki.com/heraldrywiki/wiki/Tajikistan

ORCID
Nargis Nurulla-Khojaeva ⓘ http://orcid.org/0000-0003-2178-3377

References
Abaev V.I. (1962) 'The cult of the 'seven gods' among the Scythians' [Kult 'semi bogov' u skifov] (Moscow: Ancient world) (Russian).
Acharya A. (2017) EIA Interview with Amitav Acharya on the Multiplex World Order Amitav Acharya, Adam Read-Brown, accessed 12 December 2020.
Aini S. (2005) The history of Thought Revolution in Bukhara [Ta'rihi inqilobi fikri dar Buhoro], Collected works [Kulliiet], Vol. 14. Dushanbe: Matbuot, 29–270 (Tajik)
Almond I. (2004a) The Meaning of Infinity in Sufi and Deconstructive Hermeneutics: When Is an Empty Text an Infinite One? *Journal of the American Academy of Religion*, 72:1, 97–117.
Almond I. (2004b) Sufism and Deconstruction: A Comparative Study of Derrida and Ibn Arabi (New York: Routledge).
Al-Qushayri's Epistle on Sufism (al-Risala al-qushayriyya fiilm al-tasawwuf) By Abu 'l-Qasim al-Qushayri. Transl., intr., and notes by Alexander D. Knysh. Rev. by Muhammad Eissa. (London: Garnet Publishing). 2007, xxvii + 498 pp.
Arabi I. (1980) Muhammad ibn 'Alī Ibn al-'Arabī, Ibn al-'Arabī, R. W. Austin, Issue 22 of Classics of Western spirituality Fu sû s al-hikam, The Bezels of Wisdom (Paulist Press).
Arendt H. (1958) The Human Condition. (Chicago: University of Chicago Press).
Bargués-Pedreny (2020) Resilience is 'always more' than our practices: Limits, critiques, and skepticism about international intervention, *Contemporary Security Policy*, 41:2, 263–286, DOI: 10.1080/13523260.2019.1678856

Bashier S. H. (2004) Ibn al-Arabi's Barzakh: The Concept of the Limit and the Relationship between God and the World (SUNY Press).

Bauman Z. (2001) Community: Seeking Safety in an Insecure World (Cambridge: Polity Press).

Bhabha H. (1994) The Location of Culture (New York: Routledge).

Bible Study Tools, Available at: www.biblestudytools.com/classics/the-works-of-john-donne-vol-1/sermon-xxv.html, accessed 10 November 2020

Burckhardt T. (2008) Introduction to Sufi Doctrine (World Wisdom, Inc.).

Carter M. (1998) 'Infinity and Lies in Medieval Islam,' in *Philosophy and Art in the Islamic World*, Orientalia Lovaniensia Analecta, ed. U. Vermeulen and D. De Smet, vol. 87 (Leuven: Uitgeverij Peeters).

Chandler D. (2018) 'Ontopolititcs in the Anthropocene: An Introduction to Mapping, Sensing and Hacking' (Routledge, Political Science).

Chandler D. (2020) 'Coronavirus and the End of Resilience'. Available at: www.e-ir.info/2020/03/25/opinion-coronavirus-and-the-end-of-resilience/, accessed 23 November 2020.

Chittic W. (1995) 'Ahmad Sam` n on Divine Mercy Sufi', *A Journal of Sufism*, 27: 5–11.

Chittick W. (2009) Science of the Cosmos, Science of the Soul: Pertinence of Islamic Cosmology in the Modern World (Oxford: One world).

Derrida J. (2016) Of Grammatology (Johns Hopkins University Press, Literary Criticism).

Ephraem Saint (2015) The Hymns on Faith Book collections on Project MUSE Fathers of the Church Patristic Series (CUA Press).

Flockhart T. (2016) 'The coming multi-order world', *Contemporary Security Policy*, 37:1, 3–30, DOI: 10.1080/13523260.2016.1150053

Goodman L. (1999) Jewish and Islamic Philosophy: Cross pollinations in the Classic Age (Rutgers University Press).

Hallaq W. (2015) Quranic Magna Carta: On the Origins of the Rule of Law in Islamdom. In: Magna Carta, Religion and the Rule of Law. Ed. by Robin Griffith-Jones and Mark Hill QC. (Cambridge: Cambridge University Press), 157–176.

Hallaq W. (2018) Restating Orientalism. A Critique of Modern Knowledge (Columbia University Press).

Haukkala H., van de Wetering C. and J. Vuorelma (2018) Trust in International Relations Rationalist, Constructivist, and Psychological Approaches. Routledge.

Hussain A. (2015) The Light of the Blessed Tree Islam's Intellectual Imperative in Modernity. *Doc*. Student in Islamic Studies Near Eastern Studies Department University of Michigan; Lanterna Institute.

Ibn Arabi (1980) The Bezels of Wisdom, transl. and intr. R.W.J. Austin (New York: Paulist Press).

Ibn Sina (1983) *Tahsil-us-Saodat*, Miscellany articles, Vol. 2 (Dushanbe: Irfon) [Aby Ali Ibni Sino. Tahsil Us Saodat. Osori muntahab] (Tajik).

Korosteleva E. & Flockhart T. (2020) Resilience in EU and international institutions: Redefining local ownership in a new global governance agenda, *Contemporary Security Policy*, 41:2, 153–175, DOI: 10.1080/13523260.2020.1723973

Elena A. Korosteleva & Irina Petrova (2022): What makes communities resilient in times of complexity and change? *Cambridge Review of International Affairs*, 35:2, 137–157 DOI: 10.1080/09557571.2021.2024145

Krell J. (2020) Ecocritics and Ecoskeptics: A Humanist Reading of Recent French Ecofiction Volume 5 of Studies in Modern and Contemporary France Series, (Oxford University Press).

Lotman, U.M. Text in text [Tekst v tekst], Notes of Tartu's University. Tartu, 1981. V. XIV. N 567. (Russian).

Lotman, Yuri. (2019) Culture, Memory and History: Essays in Cultural Semiotics. Ed. Marek Tamm Transl (Springer Nature).

Mahony D. (2018) 'Hannah Arendt's Ethics' (Bloomsbury Publishing).

Marx, K. 1985. 'The Civil War in France'. On the Paris Commune, by K. Marx and Fr. Engels (Moscow: Progress Publishers).

Michaeli H. (2019) Goethe's Faust and the Divan of Hafiz (Publisher: De Gruyter).

Moosa E. (2010) 'History and Normativity in Traditional Indian Muslim Thought Reading Shari'a', in the Hermeneutics of Qari Muhammad Tayyab (d. 1983). Rethinking Islamic Studies: From Orientalism to Cosmopolitanism, eds. Carl W. Ernst and Richard C. Martin; (Columbia, SC: University of South Carolina Press).

Moosa E. (2005) Ghazali and the Poetics of Imagination (Chapel Hill and London: University of North Carolina Press).

Morton T. (2013) Hyperobjects: Philosophy and Ecology after the End of the World (Minneapolis: University of Minnesota Press).

Nasr S.H. (1993) The Need for a Sacred Science (New York: SUNY Press).

Ieuan Williams (2013) Plato: All That Matters (London: John Murray)

Rumi Jalalu'l-Din (2012) Selected Poems of Rumi, Courier Corporation.

Rosenthal F. (1988) Ibn Arabi between 'Philosophy' and 'Mysticism': 'Sufism and Philosophy and Neighbours and Visit Each Other'. fa-inna at-tasawwuf wa-t-tafalsuf yatajawarani wa-yatazawarani», Oriens 31, pp. 1–35.

Saadi, R. Nazari, S. Nazari (2018) Masnawi: In Farsi with English Translation (Volume 2) (Learn Persian Online www.learnpersianonline.com), accessed 12 December 2020

Shahidi M. (2016) The Impact of the Emergence of Eurasian Art Communities in the Globalizing World. *International Relations and Diplomacy*, 4: 3, 209–218, DOI: 10.17265/2328-2134/2016.03.005

Shayegan D. (1997) Cultural Schizophrenia: Islamic Societies Confronting the West (NY: Syracuse University Press), 200.

Smirnov A.V. (2019) The All-Human and the Generally-Human [Vsechelovecheskoe vs obshchechelovecheskoe] (Moscow: Sadra: Izdatelskiy Dom YASK), 216 (Russian).

Tymieniecka A. 'The Circle of Life in Islamic Thought.' In Islamic Philosophy and Occidental Phenomenology on the Perennial Issue of Microcosm and Macrocosm. Ed. by Anna-Teresa Tymieniecka. Dordrecht: Springer, 2006, 205– 213.

William B. (2005) In the Beginning Was the Deed: Realism and Moralism in Political Argument, Princeton: Princeton University Press.

Wood R. (2017) Secrets and Aporias in Ibn al-Arabi and Derrida: A Review Essay of Ian Almond's Sufism and Deconstruction. Milestones. Commentary on the Islamic world. Available at: www.milestonesjournal.net/reviews/2017/12/27/secrets-and-aporias-in-ibn-al-arabi-and-derrida-a-review-essay-of-ian-almonds-sufism-and-deconstruction, accessed 14 October 2020.

Wuthnow R. (1989) Communities of Discourse (Cambridge, MA: Harvard University Press).

Communal self-governance as an alternative to neoliberal governance: proposing a post-development approach to EU resilience-building in Central Asia

Fabienne Bossuyt and Nazima Davletova

ABSTRACT

In the European Union's (EU) new Strategy for Central Asia, which was launched in May 2019, boosting the resilience of Central Asian societies is singled out as a key priority. Drawing on post-development thinking, this article argues that if the EU is serious about promoting resilience to empower 'the local' and contribute towards a truly sustainable future for the societies of Central Asian countries, then the EU will need to embrace a de-centred, post-neoliberal approach to resilience. This implies that the EU would have to accept 'the other' – in this case, the Central Asian societies – for what they are and advocate home-grown self-organization based on a deep understanding of the local meaning of good life and local knowledge about the available resources. Empirical illustrations to substantiate this claim are drawn from a concrete case, namely the mahalla in Uzbekistan.

Introduction

This article draws on post-development thinking to further advance the nascent scholarship that critically reviews European Union (EU) foreign policy. In recent years, a growing number of scholars have been proposing a decentred agenda for the study and practice of the EU's actorness in international relations, arguing for the need of moving away from the predominant Eurocentric views and instead adopting a decentralized standpoint (e.g., Fisher Onar and Nicolaïdis 2013; Keukeleire and Lecocq 2018). This article seeks to contribute to these scholarly debates by critically engaging with the EU's approach to resilience as part of the EU's external governance policy. Considering the limited effectiveness of the EU's promotion of governance in its neighbourhood and further afield, the article joins those scholars who argue in favour of a radical departure from the neoliberal approach that the EU follows in its conceptualization and promotion of resilience in third countries. From this perspective, the article contributes to this special issue's focus on transformations of the interaction between the EU and Central Asia by critically reflecting on how the EU's engagement with the region *should* evolve.

In the EU's latest foreign policy strategy, namely the Global Strategy, which was launched in 2016, boosting the resilience of societies in neighbouring countries and developing countries is singled out as a key priority of EU foreign policy (European Union 2016). This priority has also found its way to the EU's new Strategy for Central Asia that was launched in May 2019, in which enhancing the resilience of Central Asian societies tops the agenda (European Commission and HR 2019).

The article aligns itself with scholars who argue that if the EU is serious about promoting resilience to empower 'the local' and contribute towards a truly sustainable future for the societies of these countries, then the EU will need to embrace a de-centred, post-neoliberal approach to resilience instead of the Eurocentric, neoliberal approach that it now uses (Juncos 2017; Korosteleva 2020a,b). This implies that the EU would have to accept 'the other' – namely those societies – for what they are and advocate home-grown self-organization and self-governance based on a deep understanding of the local meaning of 'good life' and local knowledge about the available resources (Korosteleva 2020a,b).

Based on insights from post-development thinking and decoloniality, the article contributes to these academic debates by exploring local conceptualizations of the post-development concept 'good life' and local forms of self-governance and extrapolating these notions to the practice of EU foreign policy, thereby laying the foundation for a new paradigm of governance promotion that starts from local forms of self-governance that embody an indigenous understanding of good life. As community-based institutions of self-governance, which are deeply rooted in Uzbek society, especially in rural areas, the mahalla serves as a good example for illustrating how home-grown systems of self-governance that embody a local understanding of good life act in function of resilience-building. Rather than ignoring or minimizing the role that community-based systems such as the mahallas play for building resilience in local societies, the EU should endorse these governance systems and advocate them in its governance promotion efforts. However, this does not mean that the EU should be uncritical about local self-governance systems such as the mahallas. As Earle has found in the case of Kyrgyzstan, 'the uncritical promotion of "tradition" [by donors] can have negative developmental consequences, particularly for women and marginalized groups' (Earle 2005, 258). It would be naïve to think that local self-governance systems are not without their flaws and pitfalls. The mahalla is not uncontested, and has been subject to a wide range of critiques from local scholars (Dadabaev 2017; Ilkhamov 2005; Khamidova 2018; Kuliev 2019). Therefore, the EU should seek to strengthen these self-governance systems based on a deep understanding of how they function within the societal fabric of the countries concerned and of how they could be further improved.

Methodologically, the empirical illustrations draw on a combination of primary and secondary data. More specifically, they rely on insights from the literature on the role of mahallas in Uzbekistan, supplemented with findings from an original survey conducted in Uzbekistan in 2021. The survey was first conducted online nationwide (circulated in both Russian and Uzbek), followed by a physical survey conducted in the Ferghana and Andijon regions, as well as among migrant workers from those regions in Tashkent. A total of 323 respondents was reached. The online survey, which reached 240 respondents, was circulated on the basis of snowball sampling. It was first distributed among our personal networks of friends, family and colleagues in Uzbekistan, who were asked to further circulate it among their friends, family and colleagues. In order to minimize the risk of bias, which due to the online format of the survey was overwhelmingly geared towards respondents who are highly educated, young and/or based in (large) cities, it was decided to complement the online survey with a physical survey in order to reach groups of the population that were underrepresented in the online survey, namely people who are low educated, live in rural, traditional areas and/or are from an older generation. It was especially important to be able to reach people in more rural and traditional areas, as the mahallas occupy a more prominent role there. Therefore, it was decided to conduct the physical survey in the Ferghana valley, as well as among migrant workers from the Ferghana valley residing in Tashkent. Apart from Tashkent, one city in the Ferghana region, namely Margilan – which is a stronghold of conversative Islam – and five rural villages in the Andijon region (Bo'ston 1, Bo'ston 2, Yangi qadam, Xidirsha and Quruqsoy) were selected for the survey. The physical surveys, which reached 83 people (40 in Margilan, 33 in the villages and 10 in Tashkent), were conducted by two research associates who live in the respective regions in the Ferghana valley.

The article is structured as follows. It starts by demonstrating the need to radically change the EU's approach to governance promotion in third countries away from its Eurocentric, neoliberal focus to a decentred post-neoliberal perspective. Next, it advocates communal self-governance as an alternative to neoliberal governance, thereby drawing on insights from post-development thinking and decoloniality. Subsequently, the article uses the example of the mahalla in Uzbekistan to offer an entry point into how this new approach could be concretized in practice. The article concludes with some general observations and avenues for future research.

The limits of EU governance promotion in third countries

In recent years, an increasing number of scholars have been advocating a decentred agenda for the study and practice of the EU's actorness in international relations, arguing that we need to move away from the predominant Eurocentric views in the literature and instead adopt a decentralized approach (e.g., Fisher Onar and Nicolaïdis 2013; Keukeleire and Lecocq 2018). This article seeks to contribute to this agenda by critically engaging with the EU's approach to resilience. Since the launch of the EU's Global Strategy in 2016, resilience has emerged as a guiding principle of the EU's external action, and now functions as the overarching paradigm for the EU's governance promotion in third countries, in particular in the neighbourhood and the countries surrounding the neighbourhood.

A multitude of academic studies have shown the limited effectiveness of the EU's promotion of governance in the EU's neighbourhood and further afield (Börzel, Pamuk, and Stahn 2008; Delcour 2013; Freyburg et al. 2009; Hoffmann 2010; Lavenex and Schimmelfennig 2011). While having a broad understanding of governance that includes the effectiveness and inclusiveness of institutions and decision-making processes, the EU's interpretation of governance is rather political and is closely linked to the notion of good governance (Orbie et al. 2017).[1] Therefore, the EU's understanding of governance comprises elements of both 'effective governance' and 'democratic governance' (Börzel, Pamuk, and Stahn 2008; Hackenesch 2016).

When it comes to explaining the limited effectiveness of the EU's promotion of governance, most scholars adopt a rather narrow, parsimonious approach, pointing to factors such as the limited incentives offered by the EU, the deficient mechanisms and instruments that the EU relies on and its reluctance to use conditionalities due to the prioritization of interests over values (Börzel, Pamuk, and Stahn 2008; Freyburg et al. 2009; Hoffmann 2010; Lavenex and Schimmelfennig 2011). Some scholars have taken a more critical perspective and have argued that it is the EU's neoliberal approach to governance that hinders the effectiveness of the EU's governance promotion (Hout 2010; Korosteleva 2020a,b; Kurki 2011; Reynaert 2011). Looking specifically at the case of resilience-building as the EU's new paradigm to promote governance in the neighbourhood and surrounding regions spanning from Central Asia to Central Africa, this article joins those scholars who argue in favour of a radical departure from the neoliberal approach that the EU follows in its conceptualization and promotion of governance in general and resilience in particular.

Since the launch of the EU's Global Strategy in 2016, the EU has come to conceptualize resilience as 'the ability of states and societies to reform, thus withstanding and recovering from internal and external crises' (European Union 2016, 23). In what appears as a promising feature, the EU acknowledges that strengthening resilience in third countries involves granting the local societies more ownership given that 'positive change can only be home-grown' (European Union 2016, 27). Yet, despite the aim of this new resilience paradigm to balance the universalist claim of EU foreign policy (Tocci 2017, 65; Juncos 2017), in essence, the EU's understanding of and approach to resilience falls short of truly empowering the local and strengthening governance at a societal level from the bottom-up due to its continued neoliberal and Eurocentric fixation on EU norms-sharing through ready-made solutions (Joseph and Juncos 2019).

This fixation is also very clearly manifested in the EU's new Strategy for Central Asia, where the key priority of resilience is concretized by a triple focus on (1) promoting democracy, human rights and the rule of law; (2) strengthening cooperation on border management, migration and mobility, and addressing common security challenges; and (3) enhancing environmental, climate and water resilience (European Commission and HR 2019).

Joseph and Juncos explain that the EU's resilience paradigm cannot be considered entirely new and should rather be seen as 'old wine in new bottles' (Joseph and Juncos 2019, 1000). Although the EU's approach to resilience represents a move away from 'full intervention' and shifts responsibility away from the international community onto local actors,[2] thereby invoking a progressive discourse of empowerment, 'this is done according to a global template that is decided not at the grassroots level, but among international, non-governmental and donor organizations and other international actors' (999). While recognizing the leading role of partner countries, the EU is 'effectively telling them what their practices should be' (1000).

As Korosteleva (2020b, 250) explains, drawing on Joseph (2013), this neoliberal understanding of resilience externalizes Western institutions and modes of governance to local communities, 'which are then supposed to embed these solutions in the national programmes to make themselves sustainable'. In other words, the EU upholds an understanding of resilience as neoliberal governance that boils down to building resilience 'outside-in', namely 'by way of offering external solutions to internal problems for communities, turning them into dependable subjectivities and consumers of the western modes of "good governance"' (Korosteleva 2020b, 253).

This can also be observed in how the EU promotes governance and engages with local societal actors in Central Asia and how it envisions the role of civil society in these countries. Just like in other regions, the EU's approach to civil society support and governance promotion in Central Asia is neoliberal in nature, as it proceeds in a very technocratic, almost managerial manner (Axyonova and Bossuyt 2016; Keijzer and Bossuyt 2020). Moreover, the substance of its governance promotion and civil society support in Central Asia is not only embedded in the neoliberal paradigm of the state–civil society–market triangle, but also in the Western ideological concept of liberal democracy (Axyonova and Bossuyt 2016; Bossuyt and Kubicek 2015). As such, it is not surprising to see that in its engagement with local societal actors, the EU shows a continued reliance on Western-style organizations, since these have the professional systems and processes needed for accessing and managing EU funding and they better fit the EU's Western understanding of civil society (Axyonova and Bossuyt 2016; Keijzer and Bossuyt 2020).

As Korosteleva highlights, if the EU is serious about strengthening societal resilience in Central Asia as it proclaims in both the Global Strategy and the new EU Strategy for Central Asia, then it would need to decentre its approach more radically 'from those who govern to those who are subjectivized by it, and not by way of creating compliant subjects but rather by way of empowering "peoplehoods"' (Korosteleva 2020a, 3). The EU would need to reconceptualize its neoliberal, outside-in understanding of resilience and approach it instead as self-governance and self-organization based on a deep understanding of the local meaning of 'good life' (Korosteleva 2020a, 4; Korosteleva 2020b, 253). The EU would need to accept that resilience-building is about boosting 'the ability of people or a society to self-organize, drawing on its local strength and knowledge of available resources, and more importantly, on their hope for a better future' (Korosteleva and Petrova 2020, 2). This would imply going 'beyond a liberal internationalist approach of the ready-made solutions, and even beyond a new-liberal working with responsibilized subjects, from a distance' (Korosteleva 2020a, 4; see also Joseph and Juncos 2019; Rutazibwa 2014). The EU needs to acknowledge that resilience-building cannot be moulded externally and that instead it starts internally, from the communities, which draw on their existing resources and knowledge and their understanding of 'good life', with external assistance provided only as and when deemed necessary by the communities (Korosteleva 2020b, 253).

Communal self-governance as an alternative to neoliberal governance

This article aligns itself with scholars who argue that if the EU is serious about promoting resilience as a way to empower 'the local' and as a way to contribute towards a truly sustainable future for the societies of these countries, then the EU will need to embrace a decentred, post-neoliberal approach to resilience instead of the Eurocentric, neoliberal approach that it now uses (Juncos 2017; Korosteleva 2020a,b).

However, this emergent body of literature, including those studies that advocate a focus on the good life as a way to strengthen resilience (e.g., Flockhart 2020; Korosteleva and Flockhart 2020), remains heavily international relations focused and tends to rely on Western theoretical perspectives, understandings and approaches in developing their arguments, building on Western scholars such as Chandler, Weber, Bourdieu and Giddens.

To further advance this nascent body of literature, this article brings in insights from post-development thinking and decoloniality to conceptualize and operationalize a de-centred, post-neoliberal paradigm of governance promotion. If we truly want to introduce and conceptualize a de-centred approach to resilience-building, then we also need to de-centre our Western-centric conceptualizations of good life and governance, and instead start from local understandings and forms of good life and self-governance. Indeed, given that EU knowledge about 'the other' remains Eurocentric and elitist, and reproduces patterns of political, cultural and social domination, this also means that in order to enable the emergence of local conceptualizations of 'good life' and 'good society', there is a need to decolonize EU knowledge production on governance (Bridoux 2019; Rutazibwa 2014; Sadiki 2015). The idea that we need to decolonize knowledge production is strongly linked to post-development thinking. Decoloniality as a research strategy problematizes the silencing in international relations of non-Western peoples and their experiences, and sees the search for alternatives to Western-centric development frames as an integral part of knowledge production (Rutazibwa 2014, 296).

Post-development thinking has played a pivotal role in challenging the underlying assumptions and implications of classical Western development theory, as well as the subsequent neoliberal paradigm that has come to determine international development assistance and interventions. In offering viable alternatives to the neoliberal paradigm, post-development thinking has, among other things, espoused the notion of 'good life'. 'Good life' originates from the Latin American concept *Buen Vivir*, which translates as 'good living' or 'living well'. The concept emerged in Latin America as a critique of the neoliberal development strategies implemented by governments and multilateral development banks (Gudynas 2011b). By fundamentally questioning the reductionism of classical Western development theory, which reduces development to economic growth, *Buen Vivir* came to emphasize quality of life that goes beyond consumption or property by linking it to the collective well-being of humans and by seeing well-being as possible only within a community (Gudynas 2011a, 2011b). As Gudynas explains, the concept groups a set of ideas that function both as a reaction to Western development thinking – with its ideology of progress and emphasis on economic growth – and as an alternative to those conventional notions of development (Gudynas 2011b). It is in this respect that *Buen Vivir* has been linked to post-development thinking, a collection of ideas kick started by Escobar (1995) in the 1990s, and which has advocated in favour of moving beyond capitalism and the Western-centric development paradigm by promoting locally inspired alternatives to development (Schöneberg 2016).

By offering alternatives to development that emerge from indigenous traditions, *Buen Vivir* provides possibilities to move beyond the Eurocentric tradition (Gudynas 2011a). Related concepts can be found in other parts of the Global South. The Arab term *al-Harak*, or 'peoplehood', has been used to describe the local people-driven dynamics that have been unleashed across the Middle East and North Africa by the Arab Spring and captures how peoples across the region are being 'empowered to contest, redefine and reclaim a space and a voice' (Sadiki 2016, 339). Similarly, the Rwandan concept of *Agaciro*, which means 'self-dignity', is a philosophy of life and public policy in Rwanda that groups the ideals of self-determination, dignity and self-reliance (Rutazibwa 2014). *Agaciro* implies that people see themselves as 'the agents of [their] own change' and therefore draw on self-reliance based on self-knowledge (Rutazibwa 2014, 297).

By offering alternatives to development that emerge from indigenous traditions, the post-development concept of good life provides a suitable basis to move beyond the EU's Eurocentric, neoliberal approach to resilience as governance. Across the Global South, local meanings of good life represent alternatives to Western-inspired development. Importantly, these local meanings of good life have in common an emphasis on the value of self-reliance, linked to the notion that people are the agents of their own change and should act based on their local knowledge about what is good for them. This implies that these local understandings of good life and well-being are strongly reflected in local forms and practices of self-reliance and self-governance, which are key to ensuring resilience at the community level. As such, they serve as a key premise for a de-centred, post-neoliberal approach to resilience-building. In essence, this approach advocates home-grown systems and forms of self-governance and self-reliance that embody an indigenous understanding of good life. As highlighted above, this means that the EU would need to accept the local societies for what they are and acknowledge that resilience-building is about boosting the ability of local communities to self-organize based on their local strength and knowledge of available resources.

Let us now explore local meanings of good life in Central Asia and how these translate into local forms of self-governance and self-reliance that act in function of resilience-building to illustrate how this de-centred, post-neoliberal approach to resilience-building can be concretized.

The good life and communal self-governance in Central Asia

Although local meanings of 'good life' in Central Asia have not yet been studied widely in academic literature, it is possible to draw up some of the key features of the meaning of good life in the region based on existing scholarly knowledge of the region. A recent publication by European Union Central Asia Monitoring hints tentatively at what 'good life' means in present-day Central Asia, indicating the central role of the family as the main driver for a good life in the region (EUCAM 2020). Notions such as 'being part of a family, marriage, children, and contributing economically to the family' are considered important aspects of a good life (2). Almost equally important as the family in Central Asia is the community, which in some ways functions as an extended family (Kudaibergenova and Eshchanov 2020).

In a rare study of how people in Central Asia define and experience good life and well-being in direct response to neoliberal pressures, Satybaldieva (2018, 33) shows how working class people in Kyrgyzstan 'contest the neoliberal hegemony by developing alternative "caring" and "pious" selves, drawing upon traditional morality and multiple Islamic discourses to define themselves through non-market values'. Her article found that good life among the working class in Kyrgyzstan is defined as care, equality, productive labour, solidarity and economic moderation. As stated by Satybaldieva:

> Most working-class actors define themselves in relation to familial and communal ties and roles, as caring mothers, fathers and husbands and neighbours. A good life and a person of value are constituted by attachments and concerns, possessing moral sentiments that are other-oriented (such as compassion and altruism), and fostering caring relationships

with family members and neighbours. Through care and loyalty, many working class people develop a positive sense of self-worth. (42)

In Central Asia, and especially in rural areas such as the Ferghana valley, good life is closely associated with the moral principle of trust, as reflected in 'trust networks', which secure the members' good life by reciprocal practices (Boboyorov 2013, 4). This also implies that people often rely more on informal institutions and practices than on formal ones. These support networks are conditioned by reciprocity both as moral good and material aid in emergencies (Boboyorov 2013, 26). They run within families, neighbourhoods (mahallas), villages and across lines of kin, and 'involve traditional forms of community interaction, management and positions of responsibility' (Earle 2005, 249; see also Beyer and Finke 2019). As Boboyorov (2013, 4) writes, people in rural areas in southern Tajikistan trust personal networks rather than other forms of relationships. They consider personal networks 'as effective means to maintain order, certainty and security' (4). They tend to have little trust in the state institutions, whose support they see as short-term, unstable and unpredictable, and instead they rely on reciprocal and patron–client relations.

Similar to understandings of good life and well-being that can be found in other parts of the Global South (see above), the centrality of social trust and solidarity as the basis for a good life and well-being is strongly reflected in local forms and practices of self-reliance and self-governance. Earle (2005) argues that Western donors often tend to ignore these institutions, traditions and practices, as they do not fit the (Western-centric) definitions of institutions, governance and civil society.

In Kyrgyzstan and Uzbekistan, for instance, at the community-level a central role is taken up by the *aksakals*, or elders, typically well-respected older male members of the community who are given a leading role in the community, and, for example, have to help resolve local disputes (Earle 2005, 251–252). *Aksakals* are seen as symbolizing a caring civic community, which offers alternative values to the neoliberal values espoused by markets and elites (Satybaldieva 2018). At the community level, there is also the traditional practice of *khashar*, also known as *ashar*, a form of collective voluntary work, in which people from the community are expected to provide assistance for community members as part of a joint effort to improve living standards within the community (Satybaldieva 2018).

Similarly, a long-established traditional practice of self-governance in the region is the mahalla. In Uzbekistan, the mahallas continue to play a key role in offering assistance to people at the community level. The mahalla is in essence a residential neighbourhood within a village or city, which functions both as a community group based on residence and as a self-governing administrative unit (Dadabaev 2017, 78). Each mahalla is managed by a group of *aksakals*.

The extent to which local forms and practices of self-organization and self-reliance in Central Asia act in function of resilience-building has been vividly illustrated during the Covid-19 pandemic. Indeed, during the pandemic, civil society and community-based initiatives across Central Asia have been instrumental in addressing the direct implications of the pandemic, especially in areas where governments fell short, such as medical support, the provision of information and social protection (Berdiqulov, Buriev, and Marinin 2021; Cabar 2021). As such, local civil society and community-based initiatives, including self-help groups, in Central Asian countries have played a crucial role in offering life-saving assistance, not least to the most vulnerable groups in society. While the Covid-19 pandemic has revealed and exacerbated existing challenges in Central Asia relating to, among other things, poor state governance and weak state capacities, it has endorsed the key role that grassroots civil society and community-based practices of self-reliance plays in strengthening resilience in the face of a major crisis.

Nevertheless, it is important to note that home-grown community systems and practices of self-governance and self-reliance are not without their flaws and pitfalls. In particular, when it comes to the trust networks that are prevalent in rural areas in Central Asia, given that these are determined by patrimonial and patriarchal values (e.g., Roche 2020), they are not void of opportunistic behaviour on behalf of the patron and are not necessarily a panacea for promoting emancipation and overcoming social injustice. Ismailbekova (2017), for instance, shows how trust networks in Kyrgyzstan can be abused for election purposes by means of various types of election manipulation. However, for the various stakeholders in the trust network, such illegal practices are considered morally acceptable because they invest hopes and expectations of economic betterment with their 'native son' being elected to parliament. Similarly, other cases of illegal practices that are considered as morally acceptable within trust networks have been documented by Beyer (2016), such as vote-buying and corruption in the form of local politicians allocating contracts for the construction of municipal buildings. Similarly, as Satybaldieva (2018) indicates, the traditional morality – focused on loyalty and solidarity – espoused by working class people in Kyrgyzstan can result in unintended consequences, such as gendered practices and social accommodation, which represent conservative values and preclude progressive approaches to address social injustices. Similar observations have been made regarding the Mahallas (Dadabaev 2017; Ilkhamov 2005; Khamidova 2018; Kuliev 2019).

These insights alert us to what Escobar (1995, 170) and Ziai (2004, 1051) have argued from a post-development perspective, namely that one should not uncritically embrace local practices of self-reliance and self-governance as alternatives to neoliberal development. There is no such thing as 'pure [...] vernacular societies, free of domination' (Escobar 1995, 188). Therefore, to avoid romanticizing local practices of self-reliance and self-governance, decolonizing our knowledge production requires 'a critical re-evaluation of both Western and non-Western cultures, and the encounter between them' (Marglin 1990, 26). In starting from the local practices of self-governance that embody local meanings of 'good life', the proposed post-neoliberal paradigm for governance promotion should thus not remain blind to patterns of domination and abuse that undermine social justice and negatively affect societal resilience.

In what follows, the concrete case of the mahalla in Uzbekistan is offered as an example to show how the premise of the proposed post-neoliberal paradigm of governance promotion can be concretized in practice. It should be noted that the mahalla is not unique to Uzbekistan and exists also in other Central Asian countries,[3] as well as in other regions, especially in the Middle East and South Asia. In fact, in other countries, the mahalla is arguably even more prominent than in Uzbekistan. However, for a variety of reasons that go beyond the scope of this article, the mahalla in Uzbekistan tends to be one of the most documented and well-known cases, which enables us to draw on existing research. Nevertheless, while the article focuses specifically on the mahalla in Uzbekistan, the example of the mahalla more generally is well-suited for our argument, as they represent important forms of community-level governance and organization performed through community solidarity, such as religious ceremonies, life-cycle rituals (e.g., weddings and funerals), social protection provision and conflict resolution.

Mahalla: between modernity and traditionalism

The mahalla has a long history as a self-governing unit based on local identity, and it functioned as a hierarchical communal system in the pre-Soviet ages. Initially, the mahalla was part of the medieval communal management based on the legal framework of *waqf* (the Islamic equivalent of the trust) for financial administration, including tax and land management, and territorial unity of the locals. As Sievers (2002) argues, the mahallas were the only places of absolute social inclusion irrespectively of cultural, ethnic or material background. The mahallas managed to create charity-based mechanisms of resource allocation, which reached the members of the community (Sievers 2002, 109), a function that has partly remained in place up to today.

Originally, mahallas had all the features of a community-based institution, which acted as a mediator between the interests of ordinary citizens and the government, and even defended the interests of the former (Dadabaev 2017). They were also seen as a platform for bonding together and express their collective identities (Warikoo 2012). All these were relevant during pre-Soviet and even Soviet times. As some authors claim (Abramson 1998; Dadabaev 2013), the Soviet period mahallas preserved their roles as guards of collective identities partly due to the Soviet system of administration, where the former used the local mahalla structures to spread its influence. At the same time, the local population saw mahallas as a counterbalance to the foreign colonizers and showed loyalty to the institution that functioned indigenously within the traditional structures. It offered an important framework for social and practical support at times of distress (Waite 1997, 227). In Soviet-period mahallas, people used to show a voluntary corporative culture, eagerly participating in mahalla-based events, such as *khashar* and *gap*, detecting families in need and contributing willingly to weddings, funerals and other massive events of neighbours (Dadabaev 2013).

However, the term *khashar* gradually obtained a more negative connotation (e.g., Earle 2005), even being associated with forced labour. The appreciation for *khashar* mainly changed due to the transformation of socio-economic and socio-cultural relations in Uzbek society, which started in the Soviet period and continued during the years of independence. The understanding of this labour as communal and voluntary was much closer to the traditional essence of Uzbek *khashar* in the Soviet period rather than during the years of independence. In post-Soviet Uzbekistan, the activities of mahalla and *khokimiyats* (city administration) discredited themselves as pro-civilian bodies and turned into part of the state bureaucratic system in the eyes of ordinary citizens.

Political elites throughout the history of Uzbekistan have often tried to control the mahallas to enforce their own legitimacy (Abashin 2014; Ilkhamov 2005). The tendency to keep a grip on self-governing bodies in order to control the masses and monitor public sentiments strengthened in post-Soviet Uzbekistan under Karimov. The authorities used mahallas to monitor and control the public dissatisfaction (Dadabaev 2013). In turn, the mahalla has at times been seen by the people as part of the political authorities and surveillance rather than an indigenous system of social organization. Partly, the post-

Soviet mahallas took over the functions of the Soviet-era mahallas, undermining the collectivist essence of the communal activities in an attempt to pursue state propaganda. This happened mostly because the Soviet and post-Soviet mahallas were and have been heavily controlled by the state and applied a top-down approach even to those activities that previously were seen as communal.

The present-day mahalla stands on the fault line of growing traditionalism and remaining modernity in Uzbekistan. This complex phenomenon finds itself in rural and urban areas, respectively. Rural areas are still dependent on mahallas in the organization of their social life and have more trust in them. According to the field work research by Urinboyev et al., in a society where informal practices have the upper hand over legality, mahalla is still the main institution addressing the state–society relations in Uzbekistan (Urinboyev et al. 2018). In rural areas, the mahallas deal also with issues related to identity and values, whereas in the cities their work is more limited to purely legal matters. In rural areas, *mahalladoshlik* (shared mahalla origin) and mahalla-level social relations are still prominent, for example, reciprocity, trust, obligation, age hierarchies, gossip, and social sanctions (Urinboyev et al. 2018). Those ephemeral terms mean much more to people than 'official' institutions created and run by the state. The field work conducted by Urinboyev et al. even showed that mahalla leaders[4] who opt for illegal practices while negotiating with state officials over benefits for community members were much more popular than those who followed the rules and legal norms. Therefore, legal norms are usually perceived as exogenous until they are naturalized by the ordinary people and seen as corresponding to their daily needs (e.g., Ilkhamov 2005).

The mahalla as a home-grown system of self-governance

To better understand the role that the mahalla plays in Uzbekistan as home-grown systems of self-governance that embody an indigenous understanding of good life and act in function of resilience-building, a local survey was conducted. The aim of the survey was to assess to what extent the mahalla, and in particular the institutional component, is perceived among the Uzbek people as acting in function of their well-being and that of the mahalla residents. As indicated above, the survey, which consisted mainly of closed-ended questions, was first conducted online nationwide (circulated in both Russian and Uzbek), followed by a physical survey conducted in the Ferghana and Andijon regions, as well as among migrant workers from those regions in Tashkent.

Based on the answers of the Russian-speaking respondents, which represent a distinct group of the population, often considered as the 'intelligentsia' who espouse liberal values, the key components of 'a good life' are family (23.5%), financial security (18.8%) and access to education (10.6%). A total of 93% out of the 86 Russian-speaking respondents hold a university degree, and live either in the capital city of Tashkent or abroad. About 70% of those respondents replied that the mahalla institution is not important in their lives. Only 2.3% claimed the mahalla institution to be 'rather important'. They assessed the role of mahalla in their family lives as either neutral (70%) or negative (35%).

The Russian-speaking respondents perceive the social protection and provision of services as the essential functions of the mahalla (40%). However, they complained about poor delivery of these functions along with the low levels of qualifications of the mahalla staff. Responding to the open-ended questions, the respondents shared mainly negative to neutral experiences with the mahalla. They referred most frequently to the issuance of certificates by the mahalla as the only reason to rely on them, and expressed suspicions about several cases of corruption involving the mahallas. In response to the open questions, 'useless organization', 'completely corrupt' and 'harmful' were frequent descriptions of the mahalla among this group of respondents. The respondents were generally critical about the leadership of the mahallas and suggested that there is a high inclination towards corruption among them (74%) and they doubt the functionality of the management structure of the institution (81%).

These findings are in line with existing studies, which have shown that people from urban areas are more individualistic and less dependent on the institution of the mahalla, which they tend to see as a semi-state body that malfunctions and which cannot be trusted (Dadabaev 2013, 2017; Urinboyev 2011).

The Uzbek-speaking segment of respondents demonstrate a noticeable difference in attitudes than the Russian-speaking group. Although being mostly from urban areas, they represent a geographically more diverse section of the population than the Russian-speaking group, which is predominantly based in Tashkent. The Uzbek-speaking segment has a slightly different understanding of 'good life', with almost half of the respondents considering family as the key component (48.7%). A total of 20% of respondents see access to education as the main component of a good life, and for 16.7% it is financial security. The Uzbek-speaking group is more favourably disposed towards the institution of mahalla, with 44.7% and 25% considering the role of mahalla as 'slightly important' and 'very important', respectively. The Uzbek-speaking respondents tend to estimate the role of the mahalla in their family lives as 'neutral' (70%), although some see it as 'very positive' (16%). The majority of respondents assume that the mahallas are hardly independent from the state bodies and see them as poorly organized on a functional level. The Uzbek-speaking respondents assumed the public relations task of the mahalla as central (56.1%) along with the conflict resolution (almost 30%) and administrative role (19%). 42.4% of the respondents are critical about the functions of mahalla improving the lives of the residents, while 23.8% claim it to be effective in this task.

When it comes to the role of the mahalla as a custodian of traditional Uzbek values, around 30–40% in both groups of respondents replied that the modern mahallas manifest the Uzbek values accordingly. Meanwhile, 94% of the Russian-speaking respondents said the mahallas should modernize the traditional values that it exhibits, including regarding the role of women. Around 80% of the Uzbek-speaking participants expressed the same opinion, in contrast to 40% of respondents from rural areas.

These findings seem to resonate with earlier observations that suggested that the conflict between traditionalism and modernity is the strongest in urban areas. It should also be noted that the post-Soviet mahalla has been refashioned within the nation-building discourse, romanticizing the concept of the traditional structure of Uzbek society (Sievers 2002, 118). The nation-building discourse promoted by the Karimov administration embraced the mahalla as an inherent part of the traditional communal life and a custodian of Uzbek values.

The responses from the physical surveys conducted among people from the Ferghana valley seem to further corroborate insights from the literature. These respondents appear generally the most positive towards the mahalla. Among the participants of the survey conducted in Marghilan, about 60% assume the role of the mahalla in their lives to be important. This percentage is even higher among respondents from villages. However, this is not the case for migrant workers from the region who live in Tashkent, 80% of whom said that they are 'neutral' about the mahalla, and that it does not play an important role in their lives. The other 20%, in contrast, appeared to be quite positive towards the mahalla. Interestingly, these positive responses came from older participants (40 or more years old) with a medium level of education (secondary education).

The observation that the mahalla plays a more important role in rural, traditional areas than in urban areas is further evidenced in the responses concerning the main functions that the mahalla fulfils: for the majority of the respondents from the Ferghana valley, conflict resolution, services provisions such as social protection, public relations and administration are all seen as equally important tasks of the mahalla. Yet, migrant workers living in Tashkent again seem to have different views: an absolute majority of respondents (about 90%) said that the conflict-resolution role of the mahalla is the most important one, along with the services provision, mainly social protection. Moreover, while respondents from the Ferghana valley display the most positive attitude towards the Mahallas, in some cases they seem somewhat split. Among people from Marghilan, for instance, around 30% believe that the mahalla is able to change the lives of its residents for the better, while an equal number thinks the opposite.

Responses to the open questions yielded several concrete examples of how the mahalla has supported the respondents, their families or community members and contributed to their well-being. A few instances were mentioned of how the mahalla encouraged practices of *khashar* to the benefit of either the entire community or to those in need. One respondent, for example, mentioned that through *khashar* burial grounds, roads and bridges have been built in the neighbourhood. His/her mahalla also helped to secure gas supplies for the village, including for the kindergarten. Most examples that were given referred to the social protection services of the mahalla, especially financial aid to families with low incomes.

A few examples were also given of cases where the mahalla provided vital aid during the Covid-19 crisis. Others referred to the positive role that the mahalla played to resolve conflicts, whether between neighbours or in families. Yet others referred to the importance of the mahalla in organizing life-cycle rituals, such as weddings. It should be noted that also a few examples were given of how the informal mahalla has served in support of the residents. One respondent, for instance, mentioned that people from the neighbourhood are always ready to transport residents to the hospital.

However, the open questions yielded an equal number of examples of cases where the (formal) mahalla has failed to act in the interest of the respondents and/or community members. A large range of alleged cases of corruption on behalf of the mahalla staff were mentioned, mostly claiming bribery in issuing child allowances and payments to the poor. Several participants shared their discontent with the reluctance of the mahalla to assist in conflict resolution and delivery of bureaucratic help in receiving child allowances. 'The mahalla *rais* [an elected head of the mahalla committee unit] was never available when I called him for mediation in the conflict with my mother-in-law and husband,' said one respondent. The majority of female rural respondents aged 20–40 years complained about the reluctance of the mahalla to assist in receiving the child allowances or disability payments for their children. This resonates with the findings of a study commissioned by the World Bank (2019), which concluded that the formal criteria of assessment of eligibility – conducted by the mahalla – for receiving social payments can be subjectively interpreted, and that families in need may therefore get excluded.

Towards a post-development approach to EU resilience-building in Central Asia

As we have argued above, if the EU is serious about helping to boost societal resilience in Central Asia, then it needs to reconceptualize its neoliberal, outside-in understanding of resilience and approach it instead as self-governance and self-organization based on a deep understanding of the local meaning of 'good life'. The EU needs to accept that resilience-building cannot be moulded externally and that instead it starts internally, from the communities, which draw on their existing resources and knowledge and their understanding of 'good life', with external assistance provided only as and when deemed necessary by the communities.

As highlighted within post-development thinking, throughout the Global South, local meanings of good life have in common an emphasis on the value of self-reliance, linked to the notion that people are the agents of their own change and should act based on their local knowledge about what is good for them. This implies that these local understandings of good life and well-being are strongly reflected in local forms and practices of self-reliance and self-governance, which are key to ensuring resilience at the community level. This is also the case in Central Asia, where the centrality of social trust and solidarity as the basis for a good life and well-being is embedded in local forms and practices of self-reliance and self-governance, including the mahalla.

The results from our survey on the role of the mahalla serve to offer an entry point for how the EU's approach to societal resilience in Central Asia should start from local knowledge and perceptions about the potential of community-based institutions of self-governance that embody local understandings of good life to act in function of resilience-building. Rather than ignoring or minimizing the role that community-based systems and practices such as the mahallas play for building resilience in local societies, the EU should endorse these governance systems and advocate them in its governance promotion efforts as part of a decentred, post-neoliberal approach to resilience-building that truly starts from the bottom-up rather than working outside-in.

However, this does not mean that the EU should be uncritical about local self-governance systems such as the mahalla. As Earle has found in the case of Kyrgyzstan, 'the uncritical promotion of "tradition" [by donors] can have negative developmental consequences, particularly for women and marginalized groups' (Earle 2005, 258). Similarly, Beyer and Finke (2019) have clarified that the 'camouflage of factual hierarchies as performances of egalitarian ideals indicates that tradition in Central Asia can cover up or legitimize existing inequalities' (315). Therefore, the EU should seek to strengthen these self-governance systems based on a deep understanding of how they function within the societal fabric of the countries concerned and of how they could be further improved. In the case of the mahalla in Uzbekistan, this would involve accounting for the varying levels of popular support that the mahalla has among the different groups and layers of the population, which are partly connected to how the understanding of good life – and the underlying values – varies between these groups. It would also involve showing awareness of the implications of growing state involvement in such local systems in a country such as Uzbekistan, as well as of the complex and subtle balance between modernity and traditionalism, and how this balance affects people's (perceived) reliance on self-governance systems at the community level.

For instance, while the mahalla in Uzbekistan has a strong capacity to provide social protection to disadvantaged people, it nevertheless remains vulnerable to elite capture, corruption and clientelism. During the socio-economic crisis resulting from the Covid-19 lockdown, mahalla funds intended for those in need of social assistance have been mismanaged and mahallas faced numerous corruption cases (Gazeta.uz 2020). Mahalla staff have been accused of capturing public funds raised for charity purposes. Similar examples were mentioned by the survey respondents. This raises the issue of transparency since the key argument for the indigenousness of the mahalla institution is trust by the ordinary people. The discredited image of the mahallas already gradually emerged because of the strong officialization and formalization of the mahalla under Karimov. The more officialized, the less credible have become the mahalla (see also Ilkhamov 2005). Low-level bribery while issuing certificates and high-level thefts in allocation of funds have undermined trust in the mahalla over the years of independence. Still, as we have seen above, the centrality of trust networks, which can include patronage networks, means that the moral code espoused by some groups of the population considers certain illegal activities, such as bribery and vote buying, as morally acceptable (Ilkhamov 2005; Urinboyev et al. 2018).

The issue of women within the context of mahalla is another complex phenomenon, which may have controversial outcomes in the long term. Putting women under the narrative of traditional settings and addressing the gender-related issues within the context of family and state ideology of ma'naviyat vas ma'rifat (spirituality and enlightenment) are realized through the official mahalla. Thus, mahallas can be considered responsible for forcing women to stay in abusive marriages to keep the number of divorces in the community low. The reconciliation committees under the mahalla, where female leaders reside, in majority cases apply pressure on women appealing to traditions and implying victim-blaming in an attempt to preserve families (Davletova 2020).

As Waite (1997, 227–228) argues, 'the mahalla environment may be oppressive to those who do not wish to conform with the generally conservative moral codes it promotes'. The moral codes in a patriarchal society are more easily applied to women, which is the case in Uzbekistan. If in the Soviet propaganda women were represented as part of the secular emancipation within the Communist ideology, independent Uzbekistan made women part of the nation-building narrative with very limited roles of mother and wife (Davletova 2020). This shows that the EU would have to be careful not to buy into the 'myth of the community' and thus not romanticize the mahallas as an ideal of home-grown self-governance. The EU should promote the mahallas, but not without 'an attempt to examine whether these traditional practices or positions of authority are, as far as possible, accountable to or legitimized by the majority of community members' (Earle 2005, 258).

Conclusions

The article has sought to contribute to the emergent scholarship that argues that if the EU is serious about promoting resilience to empower 'the local' and contribute towards a truly sustainable future for the societies of Central Asian countries, then the EU will need to embrace a de-centred, post-neoliberal approach to resilience instead of the Euro-centric, neoliberal approach that it now uses. This means that the EU would have to accept those societies – for what they are and advocate home-grown self-organization and self-governance, starting from a deep understanding of the local meaning of 'good life' and local knowledge about the available resources. Drawing on insights from post-development thinking and decoloniality, the article explored local conceptualizations of 'good life' and local forms of self-governance with a view to extrapolating these notions to the practice of EU foreign policy, thereby laying the foundation for a new para-digm of governance promotion that starts from the local forms of self-governance that embody an indigenous understanding of good life. As community-based institutions of self-governance, which are deeply rooted in Uzbek society, especially in rural areas, the mahalla in Uzbekistan has served as a concrete empirical illustration of how the premise of this post-neoliberal paradigm of governance promotion could be operationa-lized. We have argued that rather than ignoring or minimizing the role that community-based systems such as the mahalla play for building resilience in local societies, the EU should endorse these governance systems and advocate them in its governance pro-motion efforts. The results from our survey have provided concrete insights into how people in Uzbekistan perceive the mahalla to be acting in function of local resilience-building.

However, the significant variation in attitudes towards the mahalla among the groups of respondents, which are in part linked to how the understanding of good life – and the underlying values – differs between these groups, has vividly indicated that the EU should not be uncritical about local self-governance systems such as the mahalla. Instead, the EU should seek to support these self-governance systems and practices based on a deep understanding of how they function within the societal fabric of the countries concerned and of how they could be further improved. Clearly, the latter is unlikely to be an easy task, especially if the EU is to avoid interference in local affairs, with the potentially adverse effect of aggravating local societal cleavages. Therefore, for the EU to move towards an effective inside-out approach to societal resilience, it will be essential for the EU to only help as and when requested by the communities and to proceed only if the respective local practices of self-governance and self-organization, and their positions of authority, are accountable to or legitimized by the majority of community members.

The present article offers only a first entry point into how the EU's approach to societal resilience in Central Asia should start from local knowledge and local perceptions about the potential of community-based forms of self-governance that embody local under-standings of good life to act in function of resilience-building. Future research could explore this avenue in more depth and provide other empirical illustrations to further con-cretize this new paradigm for EU governance promotion. Moreover, particular attention should be paid to more informal practices of communal self-reliance. One of the limit-ations of our research is that it focused mainly on the mahalla as a formal unit of self-gov-ernance and largely ignored informal practices of self-reliance at the mahalla level, which are equally, if not more, relevant than the formal ones considering that people in Central Asia often rely more on informal institutions and practices than on formal ones.

Notes

1. As operationalized by the EU for its development policy in 2004, the EU's understanding of good governance includes six components: democratization and elections; the promotion and protection of human rights; strengthening of the rule of law; enhancement of the role of civil society; the reform of public administration, the civil service and public finance management; and decentralization and capacity-building of local government (EuropeAid Cooperation Office 2004).
2. The EU's resilience approach is resonant of a similar trend among several international organizations, including the United Nations (UN), the Organization for Economic Co-operation and Development (OECD) and the World Bank. In fact, in operationalizing its approach, the EU even draws inspiration from the UN (Joseph and Juncos 2019, 1000).
3. For the role of the Mahallas in Tajikistan, see, for example, Boboyorov (2013) and Cieślewska (2015).
4. Mahalla committees are headed by senior local residents who are nominated and paid for by the state. According to Waite (1997, 227), they are 'likely to have a degree of legitimacy in [their] own community, derived from traditional respect for elders, and in many places, growing links with the local mosque'.

Disclosure statement

No potential conflict of interest was reported by the authors.

References

Abashin, S. 2014. "Советский кишлак: между колониализмом и модернизацией." Новое литературное обозрение.

Abramson, D. 1998. "From Soviet to Mahalla: Community and Transition in Post-Soviet Uzbekistan." *PhD thesis, Indiana University*. Ann Arbor: UMI Dissertation Services.

Axyonova, V., and F. Bossuyt. 2016. "Mapping the Substance of the EU's Civil Society Support in Central Asia: From Neo-Liberal to State-Led Civil Society." *Communist and Post-Communist Studies* 49 (3): 207–217.

Berdiqulov, A., M. Buriev, and S. Marinin. 2021. *Civil Society and the COVID-19 Governance Crisis in Kyrgyzstan and Tajikistan*. Berlin: Institut für Europäische Politik.

Beyer, J. 2016. *The Force of Custom Law and the Ordering of Everyday Life in Kyrgyzstan*. Pittsburgh: University of Pittsburgh Press.

Beyer, J., and P. Finke. 2019. "Practices of Traditionalization in Central Asia." *Central Asian Survey* 38 (3): 310–328.

Boboyorov, H. 2013. "The Ontological Sources of Political Stability and Economy: Mahalla Mediation in the Rural Communities of Southern Tajikistan". Bonn: Crossroads Asia Working Paper Series, 13.

Börzel, T. A., Y. Pamuk, and A. Stahn. 2008. "The European Union and the promotion of good governance in its near abroad. One size fits all?". *Working Paper* 18, SFB-Governance, Berlin.

Bossuyt, F., and P. Kubicek. 2015. "Favouring Leaders Over Laggards : Kazakhstan and Kyrgyzstan." In *The Substance of EU Democracy Promotion : Concepts and Cases*, edited by A. Wetzel, and J. Orbie, 177–192. Basingstoke: Palgrave Macmillan.

Bridoux, J. 2019. "Shaking off the Neoliberal Shackles: "democratic Emergence" and the Negotiation of Democratic Knowledge in the Middle East North Africa Context." *Democratization* 26 (5): 796–814.

Cabar. 2021. "State and civil society during the COVID-19 pandemic in Central Asia", 27 September, Retrieved 29 December 2021, from https://cabar.asia/en/state-and-civil-society-during-the-covid-19-pandemic-in-central-asia.

Cieślewska, A. 2015. *Community, the State and Development Assistance: Transforming the Mahalla in Tajikistan.* Księgarnia Akademicka: Kraków.

Dadabaev, T. 2013. "Community Life, Memory and a Changing Nature of Mahalla Identity in Uzbekistan." *Journal of Eurasian Studies* 4 (2): 181–196.

Dadabaev, T. 2017. "Between State and Society: The Position of the Mahalla in Uzbekistan." In *Social Capital Construction and Governance in Central Asia. Politics and History in Central Asia*, edited by T. Dadabaev, M. Ismailov, and Y. Tsujinaka, 77–95. New York: Palgrave Macmillan.

Davletova, N. 2020. "Women of Uzbekistan: Empowered on Paper, Inferior on the Ground." *New Voices from Central Asia*, edited by M. Laruelle, 124–139. Societal Transformations.

Delcour, L. 2013. "Meandering Europeanisation. EU Policy Instruments and Policy Convergence in Georgia Under the Eastern Partnership." *East European Politics* 29 (3): 344–357.

Earle, L. 2005. "Community Development, 'Tradition' and the Civil Society Strengthening Agenda in Central Asia." *Central Asian Survey* 24 (3): 245–260.

Escobar, A. 1995. *Encountering Development. The Making and Unmaking of the Third World.* Princeton: Princeton University Press.

EuropeAid Cooperation Office. 2004. "Handbook on Promoting Good Governance in EC Development and Co-operation." In *EC Development and Co-operation.* Brussels: European Commission.

EUCAM. 2020. What does a 'good life' mean for Central Asia? *EUCAM Watch*, No. 22.

European Commission and High Representative of the Union for Foreign Affairs and Security Policy. 2019. Joint Communication on the EU and Central Asia: New opportunities for a stronger partnership, Brussels, 15 May. Retrieved 14 June 2019, from https://eeas.europa.eu/headquarters/headquartershomepage/62411/european-union-and-central-asia-new-opportunities-stronger-partnership_en.

European Union. 2016. Global Strategy for the European Union's Foreign and Security Policy: 'Shared Vision, Common Action: A Stronger Europe', Brussels, June 2016.

Fisher Onar, N., and K. Nicolaïdis. 2013. "The Decentring Agenda: Europe as a Post-Colonial Power." *Cooperation and Conflict* 48 (2): 283–303.

Flockhart, T. 2020. "Is This the End? Resilience, Ontological Security, and the Crisis of the Liberal International Order." *Contemporary Security Policy* 41 (2): 215–240.

Freyburg, T., S. Lavenex, F. Schimmelfennig, T. Skripka, and A. Wetzel. 2009. "EU Promotion of Democratic Governance in the Neighbourhood." *Journal of European Public Policy* 16 (6): 916–934.

Gazeta.uz. 2020. *Выявлено хищение средств из фонда «Махалла» в Сергели.* July 7. Accessed July 2020. https://www.gazeta.uz/ru/2020/07/07/money/.

Gudynas, E. 2011a. *"Buen Vivir*: Today's Tomorrow." *Development* 54: 441–447.

Gudynas, E. 2011b. "Good Life: Germinating Alternatives to Development." *America Latina en Moviemento*, issue 462, February 2011.

Hackenesch, C. 2016. *Good Governance in EU External Relations: What Role for Development Policy in a Changing International Context?* Study prepared for the Policy Department of the Directorate-General for External Policies of the European Parliament, EP/EXPO/B/DEVE/2015/02 EN.

Hoffmann, K. 2010. "The EU in Central Asia: Successful Good Governance Promotion?" *Third World Quarterly* 31 (1): 87–103.

Hout, W. 2010. "Governance and Development: Changing EU Policies." *Third World Quarterly* 31 (1): 1–12.

Ilkhamov, A. 2005. "The Thorny Path of Civil Society in Uzbekistan." *Central Asian Survey* 24 (3): 297–317.

Ismailbekova, A. 2017. *Blood Ties and the Native son. Poetics of Patronage in Kyrgyzstan.* Bloomington, IN: Indiana University Press.

Joseph, J., and A. E. Juncos. 2019. "Resilience as an Emergent European Project? The EU's Place in the Resilience Turn." *Journal of Common Market Studies* 57 (5): 995–1011.

Joseph, J. 2013. "Resilience as Embedded Neoliberalism: A Governmentality Approach." *Resilience* 1: 38–52.

Juncos, A. E. 2017. "Resilience as the new EU Foreign Policy Paradigm: A Pragmatist Turn?" *European Security* 26 (1): 1–18.

Keijzer, N., and F. Bossuyt. 2020. "Partnership on Paper, Pragmatism on the Ground: The European Union's Engagement with Civil Society Organisations." *Development in Practice* 30 (6): 784–794.

Keukeleire, S., and S. Lecocq. 2018. "Operationalising the Decentring Agenda: Analysing European Foreign Policy in a Non-European and Post-Western World." *Cooperation and Conflict* 53 (2): 277–295.

Khamidova, S. 2018. "Civil society in Uzbekistan: Building from the ground up." *EUCAM Commentary*, No. 31, July 2018.

Korosteleva, E. 2020a. "Paradigmatic or Critical? Resilience as a New Turn in EU Governance for the Neighbourhood." *Journal of International Relations and Development*, doi:10.1057/s41268-018-0155-z.

Korosteleva, E. 2020b. "Reclaiming Resilience Back: A Local Turn in EU External Governance." *Contemporary Security Policy* 41 (2): 241–262.

Korosteleva, E., and T. Flockhart. 2020. "Resilience in EU and International Institutions: Redefining Local Ownership in a new Global Governance Agenda." *Contemporary Security Policy* 41 (2): 153–175.

Korosteleva, E., and I. Petrova. 2020. "Resilience Is Dead. Long Live Resilience?" Italian International Affairs Institute, https://www.iai.it/en/pubblicazioni/resilience-dead-long-live-resilience?fbclid=IwAR3PTlugF1wKU0qJtkzaoJFxziDqVpEMdOU7roRAin3dD9uFJEU6HL9y1WQ.

Kudaibergenova, D., and B. Eshchanov. 2020. "What Does 'Community' Mean? Family, Social Trust and Everyday Life in Uzbekistan", draft paper presented at the COMPASS SI Workshop, 18 June.

Kuliev, K. 2019. Uzbekistan: Why Should the State Weaken Control Over the Institute of Makhalla? *Central Asian Bureau for Analytical Reporting*, 17.01.2019, online at https://cabar.asia/en/why-should-the-state-weaken-control-over-the-institute-of-makhalla/.

Kurki, M. 2011. "Governmentality and EU Democracy Promotion: The European Instrument for Democracy and Human Rights and the Construction of Democratic Civil Societies." *International Political Sociology* 1 (5): 349–366.

Lavenex, S., and F. Schimmelfennig. 2011. "EU Democracy Promotion in the Neighbourhood: From Leverage to Governance?" *Democratization* 18 (4): 885–909.

Marglin, S. 1990. "Towards the Decolonization of the Mind." In *Dominating Knowledge: Development, Culture and Resistance*, edited by F. Apffel-Marglin, and S. Marglin, 1–28. Oxford: Clarendon.

Orbie, J., F. Bossuyt, P. Debusscher, K. Del Biondo, S. Delputte, V. Reynaert, and J. Verschaeve. 2017. "The Normative Distinctiveness of the European Union in International Development : Stepping out of the Shadow of the World Bank?" *Development Policy Review* 35 (4): 493–511.

Roche, S. 2020. "The Family in Central Asia: New Research Perspectives." In *The Family in Central Asia: New Perspectives*, edited by S. Roche, 7–39. Berlin: De Gruyter.

Reynaert, V. 2011. "Preoccupied with the Market: The EU as a Promoter of 'Shallow' Democracy in the Mediterranean." *European Foreign Affairs Review* 16 (5): 623–637.

Rutazibwa, O. U. 2014. "Studying *Agaciro*: Moving Beyond Wilsonian Interventionist Knowledge Production on Rwanda." *Journal of Intervention and Statebuilding* 8 (4): 291–302.

Sadiki, L. 2015. "Towards a 'Democratic Knowledge' Turn? Knowledge Production in the Age of the Arab Spring." *Journal of North African Studies* 20 (5): 702–721.

Sadiki, L. 2016. "The Arab Spring: The 'People' in International Relations." In *International Relations of the Middle East*, edited by L. Fawcett, 325–355. Oxford: Oxford University Press.

Satybaldieva, E. 2018. "Working Class Subjectivities and Neoliberalisation in Kyrgyzstan: Developing Alternative Moral Selves." *International Journal of Politics, Culture and Society* 31: 31–47.

Schöneberg, J. 2016. *Making Development Political. NGOs as Agents for Alternatives to Development.* Nomos: Baden-Baden.

Sievers, E. W. 2002. "Uzbekistan's Mahalla: From Soviet to Absolutist Residential Community Associations." *The Journal of International and Comparative Law at Chicago-Kent* 2: 91–155.

Tocci, N. 2017. *Framing the EU Global Strategy: A Stronger Europe in a Fragile World*. Basingstoke: Palgrave.

Urinboyev, R. 2011. "Bridging the State and Society: Case Study of Mahalla Institutions in Uzbekistan." in Norms between Law and Society: A Collection of Essays from Doctoral Candidates from Different Academic Subjects and Different Parts of the World, 115–133.

Urinboyev, R., A. Polese, M. Svensson, L. Adams, and T. Kerikmae. 2018. "Political vs Everyday Forms of Governance in Uzbekistan: The Illegal, Immoral and Illegitimate Politics and Legitimacy in Post-Soviet Eurasia." *Studies of Transition States and Societies* 10 (1): 50–62.

Waite, M. 1997. "The Role of the Voluntary Sector in Supporting Living Standards in Central Asia." In *Household Welfare in Central Asia*, edited by J. Falkingham, J. Klugman, S. Marnie, and J. Micklewright, 221–235. London: Palgrave Macmillan.

Warikoo, K. 2012. "Tradition and Modernity in Uzbekistan. The Role of Mahalla in Local Self-Governance." *Himalayan and Central Asian Studies* 16 (3/4): 31.

World Bank. 2019. *Uzbekistan-Social Assistance Targeting Assessment*. Social Assistance Targeting Assessment, 1–38. Washington, DC: World Bank.

Ziai, A. 2004. "The Ambivalence of Post-Development: Between Reactionary Populism and Radical Democracy." *Third World Quarterly* 25 (6): 1045–1060.

Belarus between West and East: experience of social integration via inclusive resilience

Victor Pravdivets (iD), Anna Markovich (iD) and Artsiom Nazaranka (iD)

Abstract   *The article explores aspects of resilience developed in Belarusian society and dwells on those elements of resilience as viewed by the general public in 2019. Based on the results of focus groups conducted in Belarus, the paper uncovers perceptions of identity and community of relations, explores the Belarusian vision of a good life, and substantiates the importance of value orientations as an element of social integration. The article expands the understanding of resilience, introducing the notion of inclusive resilience, thus highlighting the importance of local coping strategies while addressing global challenges.*

Introduction

Resilience-thinking is an increasingly influential approach to understand and grapple with the peculiarities of functioning modern societies that are affected by crises, uncertainty and change. Theoretical approaches to the concept of resilience have been developed in several studies (Chandler 2014; Bourbeau 2015) that analyze the concept and explain it at a local level. In particular, a critical turn in resilience studies, focusing on 'the local' and 'the person' has been recently developed by Chandler (2020), and Korosteleva and Flockhart (2020). They examine governing regimes of international institutions and unpack local sources of resilience, both conceptually and empirically. The notion of resilience is further elaborated in the introduction and contributions to this volume, identifying resilience's components and variations, such as inclusive resilience and variants of resilience manifested in the post-soviet Muslim communities. Thus, the view of resilience both as a quality of a complex system that allows it to return to optimum functionality after a crisis and as a self-governance strategy underlies this paper. Using this definition, the authors attempt to explore how resilience is perceived by the representatives of the general population in Belarus, analysing first-hand data of focus groups. They discuss the Belarusian society and resilience-building, and in doing so consider participants' forms of identity, interpretations of 'good life', discussing the existing support infrastructures (family, friends, neighbourhood, the state), and visions of the country's international cooperation directions. The inquiry of resilience via its core constitutive elements introduces a type of

The work is prepared in the framework of the Global Challenges Research funded project ES/P010849/1 'Comprehensive capacity-building in Eastern Neighbourhood and Central Asia: research integration, impact governance and sustainable communities' (GCRF COMPASS).

'inclusive resilience' that is characteristic of behavioural strategies articulated at the grassroots level in contemporary Belarusian society.

This study is premised on the data collected during May - June 2019. The information analyzed below is important as it provides an opportunity to consider views by representatives of Belarusian society a year before the turbulent events of 2020, and to study the point that preceded the period of escalation of social tensions in Belarus. Specifically, six focus groups (FG) were conducted in all the administrative district centres of Belarus. Each group involved from eight to eleven participants. In total, fifty-four people took part in the FGs, representing all the of age socio-demographic groups (by gender, age and level of education).

The obtained project data make it possible to examine aspects of contemporary Belarusian society, such as identity, sense of belonging, social capital and values, life satisfaction, perception of 'good life', pressing issues, sense of security, social well-being, country image and national positioning. These aspects assist in viewing and analysing resilience as a quality of a complex (adaptive) system that enables communities (nations here) to more adequately respond to change in search of equilibrium (see the introduction to this volume) and serve as a new vision of governance for an increasingly complicated world (Korosteleva 2020).

The article starts with a short presentation of the technical details of the research, describing the essential features of the conducted focus groups. Further on, the paper enquires into the concepts of national identity as viewed by the focus-group participants and investigates their visions of 'good life', proceeds with the presentation of the historical community of relations and introduces the contemporary community of relations as a present-day source of local resilience. The participants' ideas on the resources of strength and sustainability of their country in the international context provide essential data for the interpretation of their ideas of national resilience in the global world.

Thus, based on the first-hand data, the work expands resilience theory, unpacking the 'local' sources of resilience, providing meaningful insights into the resilience concepts at the 'local' level. Simultaneously, the analysis contributes to a detailed comprehension of the nature and peculiarities of the Belarusian version of resilience.

Methodology

Focus group discussions were the method of data collection. They were conducted in all the district centres and the capital of Belarus. The participants of the group discussions were recruited through the 'snowball' method. The respondents participating in the focus groups were selected by predetermined quotas. The quota characteristics were gender, age and level of education.

Focus group moderators were trained sociologists with university degrees who had previous experience of moderating focus groups. They participated in a training session for moderators, organised in the frameworks of the study, to discuss and comprehend the peculiarities of the project research objectives, the aims of the focus groups and the features of the guide.

During the period from May to June 2019, the trained moderators conducted six focus groups—one focus group discussion per district centre and

№	Date	City	Number of respondents	Length
1	21 May 2019	Mogilev	10	2 h 35 min
2	23 May 2019	Minsk	11	2 h 15 min
3	30 May 2019	Grodno	9	2 h 00 min
4	21 June 2019	Gomel	8	1 h 30 min
5	21 June 2019	Brest	8	1 h 50 min
6	24 June 2019	Vitebsk	8	1 h 40 min

one in the capital of the Republic of Belarus. The table below presents the details of the carried-out focus groups:

In total, fifty-four respondents participated in the focus groups. The participants represented various social-demographic strata. They included twenty-seven males and twenty-seven females. The age of the respondents varied from eighteen to seventy years. In particular, three groups of respondents were selected: twenty respondents aged from eighteen to thirty-five, seventeen respondents aged from thirty-six to fifty-four, seventeen respondents over fifty-five years of age. The longest focus group lasted for two hours and thirty-five minutes. The shortest one took one hour and a half. The average focus group length was one hour and fifty-eight minutes.

The respondents gave their informed consent to be video-recorded in the course of the group discussions and were provided with guarantees to safeguard their confidentiality. The focus group material was verbatim transcribed, presenting all the questions, comments and answers. The verbatim transcripts do not contain the personal data of the respondents to prevent any opportunity to identify the participants.

The respondents demonstrated active participation, expressed their opinions on the questions and issues discussed during the discussions, despite the duration of some of the focus groups.

The respondents were asked about their feeling of self-identification and belonging to social groups. They shared who they reached out to in case of problems or challenges. They explained their visions of 'good life' and its aspects. In addition, the participants explained their main values and named what they valued most of all in their national or ethnic identity. The moderators took interest in the challenges the country faced at that moment. Furthermore, they asked who their country could rely on to make the country better, or in case of troubles and hardships. The final set of questions concerned visions of the future: the heritage the participants would like to pass on to future generations.

All the issues mentioned above were discussed with the participants to gain their first-hand interpretations of the concepts that constitute the ability of a population to survive in time of hardships. Furthermore, the data provide an opportunity to get a more detailed comprehension of the nature and peculiarities of the local version of resilience.

Concepts of self-identification and territorial reference as a basis for social inclusion

In trying to untangle resilience, one has to view the concepts and processes that make communities transform and endure in the face of troubles. The constitutive categories of this study are identity, visions of 'good life', social structures of communal support as manifested in participants' interpretations.

The concept of self-identification is essential for comprehending resilience because self-identification is a core component of personality. The widely used notions of identity by the focus group participants are such typical concepts as profession: 'I am a driver' (male, thirty-six, Minsk); nationality: 'I am a Belarusian...' (female, thirty-six, Mogilev); citizenship: 'I am a citizen of the Republic of Belarus' (male, twenty-two, Grodno). Also, the respondents identify themselves as belonging to certain social groups: 'I am a working pensioner' (female, sixty-five, Gomel); their family: 'I am a mother of two children, marital status - married, my husband works in business' (female, thirty-three, Brest); hobby groups: 'In particular, I can refer myself to the group of gamers' (male, twenty-two, Grodno). Such concepts look ordinary. In addition, the FG participants of Belarusian origin identify themselves via the location of birth in Belarus and the location of their residence: 'Me, too, I am from Belarus, a native from Ushachi area, I grew up in Vitebsk region' (female, forty-four, Vitebsk); 'I am a native resident of Gomel' (male, sixty-three, Gomel).

An important determinant in the self-identification of the citizens of Belarus of non-Belarusian origin is also territoriality regarding the country as a whole. For many, the fact of birth or even long-term residence in Belarus, regardless of their ethnic origin, is a reason to consider oneself to be a Belarusian. The respondents refer to the long period of residence in Belarus as one of the significant grounds when identifying themselves as the Belarusians: 'When introducing myself, I've told that in my family there are a lot of nationalities. The grandparents. My wife is not a Belarusian, but we live in Belarus. So, the relatives, everyone considers oneself to be a Belarusian. Ask my wife, who she is. She will tell: "I am a Belarusian," although she is not Belarusian' (male, forty-one, Minsk); 'If we live in this territory, if we have adopted these traditions, customs, we like to live here, why not consider ourselves as Belarusians? Even if you are Russian by blood' (male, thirty-five, Mogilev). As one can see, the inhabitants of Belarus can identify representatives of the other ethnic groups residing in the country as the Belarusians, if they have been living at this land for long, complying with the generally accepted way of life here.

Thus, along with reference to professional and social groups, locality and territory, nationality and citizenship, the respondents can relate their identity to long-term residence in Belarus joined with respect to its culture, values, traditions. Therefore, one can state a parallel existence of a variety of identities, based on various grounds, where the civic type of identity is the most embracing.

Describing the qualities of the Belarusians, the focus group participants talk about their tolerance, hard work, amiability and peacefulness, a predisposition to compromise in potential conflict situations. For instance, several focus group participants note a relatively high tolerance levels in Belarusians: 'To a greater extent, maybe, by ninety-nine percent, I believe that our citizens are tolerant...' (male, twenty-three, Grodno). The focus here is on the people of the other ethnic groups, migrants, as well as those holding a different faith. However, it excludes the cases, when the behaviour of the others comes into conflict with the established norms and values accepted in Belarus: 'On the other hand, in our country, it is customary to accept anyone. And those refugees... And to find something to do for them... If a person is ready to work for the good of our republic, maybe it's a good thing' (female, forty-five, Vitebsk). The experience of regulation of interethnic relations during the period when Belarus was a part of the USSR was also mentioned by the FG participants among the reasons for a tolerant attitude towards the representatives of other ethnic groups.

The participants mentioned the absence of the desire for greatness as a feature of the Belarusians: 'We don't need this scale. We need to earn a penny, spend it quietly, and relax' (male, forty-four, Minsk); 'I don't want to live in a "great" country at all. I want to live in a small or medium-sized, comfortable, clean, safe country' (male, thirty-eight, Vitebsk).

It correlates with another important quality that characterizes the Belarusian identity, according to many participants in the focus groups, that is the desire for stability: 'Stability can be called the most important value, that is the state when everything is good. This is the most important value...' (male, twenty-two, Mogilev). Stability is a component of 'good life' for them: 'Stability in the family, at work, to have no debts, to have a stable income' (female, fifty-one, Vitebsk). It is often mentioned together with the desire to 'live your own little quiet life' and the wish to avoid any changes even at the elementary daily level. Although, there are single participants who view stability as a sign of stagnation: 'You can't fool us with the word "stability." It has long been called the right word for "stagnation"' (male, sixty-five, Gomel). Though, for most of the participants, stability is an appreciated value, equal to security.

The participants mention family, health, good relations with people among the other values: 'For me, the greatest value is family, first and foremost' (male, twenty-three, Grodno); 'Good relationships with people are also a value. Health is a value for me' (male, fifty-five, Mogilev).

Belarusian national history is full of political, social, economic transformations. In particular, the country's geographical location made it necessary for the local people to be able to live in peace with representatives of various ethnic groups. At present, being a multinational country—dozens of ethnic groups live in the republic, according to the national census of 2019 (*Statistical yearbook 2020*, 47)—Belarus turns out to be one of a few post-Soviet states that managed to avoid serious ethnic conflicts after the disruption of the Soviet Union.

According to the 2019 census, 84.9% of the residents of Belarus refer to themselves as Belarusians. Only 7.5% of the residents of the republic ethnically determine themselves as the Russians (2020, 47). Along with that, 69.8% of those calling themselves Belarusians use the Russian language in daily communication (*Statistical yearbook* 2020). Although the Belarusian language is considered to be the native one by 61.2% of the ones who ethnically consider themselves to be the Belarusians, only 28.5% of the Belarusians speak the language at home (Hentschel and Zeller 2014). This determined the current situation where the status of the state language is assigned to both Belarusian and Russian in the legislation of the republic. On the one hand, this helps prevent a possible split in the Belarusian society, enables representatives of the main linguistic groups to feel fully integrated into society, aids minor linguistic groups, for instance, the Ukrainians and Poles to facilitate their interactions in the country. On the other hand, according to the data of this study, some Belarusians are worried that the Belarusian language has been losing its popularity and relevance in society over time.

The respondents consider the knowledge of two state languages to be an obligatory condition for being a Belarusian: 'To make your future better, first of all. If you live in Belarus, and you identify yourself as a Belarusian, you must know both the state languages' (male, forty-five, Vitebsk). Also, the Belarusian language is mentioned as a heritage that the FG participants would like to pass on to future generations: 'Of course, the mission of this generation is, firstly, to keep the Belarusian language alive so that future generations can use it to communicate' (female, thirty-five, Brest). Some of the respondents are worried that the gradual decline in its use is taking place at the moment, and as a result the Belarusian language has become unclaimed: 'The Belarusian language, unfortunately, is on the verge of survival and extinction. I don't know how to say it exactly. It barely lives' (female, thirty-nine, Minsk). Simultaneously, there are distinct voices, telling that coupling two national languages at the official level was a wise decision taken in the past. Moreover, it is still claimed by the population in the contemporary conditions: 'Also I would like to add, as far as the language is concerned. Do you remember, I guess, the nineties, the Belarusisation? We had such a trend. But, still, a reasonable decision was made to combine the Russian language with the Belarusian one. Because we see the example of Ukraine today: what happened and what it led to - the division of society' (female, forty-eight, Grodno).

One can see that the respondents would not like the Belarusian language to disappear. However, they do not contribute to it and do not speak the language. It means they do not invest in its preservation: 'We have to be. But why are we - the Belarusians - not the native speakers of the Belarusian language?' (male, twenty-three, Grodno). Therefore, despite the participants' complaining attitudes, it is possible to state that they are passive in addressing the linguistic situation.

Frequently, language is a meaningful element of national and ethnic identification. The expressed anxiety about the linguistic situation is a collateral indication of the possible interest towards ethnic identity, on the one hand. On the other hand, the participants do not express any readiness to make significant changes in this respect. Along with that, the respondents assert their national identity with reference to their state clearly and confidently.

The concept of 'peoplehood' and 'tutejshyja'

The described reference to territoriality above requires a brief consideration of the concept of 'peoplehood' as presented in the Belarusian culture. For instance, the study of the verbal expression of the major lingua-mental units (concepts) presented in the Belarusian literary texts models the nominative field of the concept 'peoplehood'. It demonstrates that the lexeme 'tutejshyja' is one of the linguistic units representing the concept of 'peoplehood' (Pivavar 2015, 145). The word 'tutejshyja' can be translated in English as 'the locals'. One FG respondent also mentions the locals as 'tuteishyja', when recalling the old people, the Belarusian history: 'A lot of old people, who passed away already. They were the "tutejshyja"—either the Poles or the Belarusians' (male, thirty-eight, Vitebsk). Considering that the respondent was speaking Russian at the moment of the focus-group, he, nevertheless, used the word in Belarusian. It indicates that it is semantically meaningful for him to distinguish the representatives of the Belarusian people of the first half of the twentieth century as 'tutejshyja'. The respondent interprets 'tutejshyja' as 'either the Poles, or the Belarusians'. It is important to stress, that at present the respondents call themselves the Belarusians without any hesitation: 'I'm Belarusian. I feel a sense of belonging to my country' (female, thirty-six, Mogilev). It is supported by the study of the key concepts in the logo-sphere of the Belarusian culture as represented in the Belarusian print media (Samusevich 2012). One of the key concepts is the concept of 'people'. The slot 'tuteishyja' was present and active in the Belarusian print media in the beginning of the twentieth century (2012, 47). However, by the beginning of the twenty-first century the concept 'people' does not contain 'tuteishyja' at all as presented in the print media (2012, 97). Thus, it refers to Belarusian history only.

All in all, the territorial references in self-identification are either very specific or very general. On the one hand, the specific references can indicate the area or the town of origin of a Belarusian to confirm it clearly for the other Belarusians. The general reference, on the other hand, embraces the country. Along with the ethnic Belarusians, it can incorporate the people who are aware that ethnically they are not of Belarusian origin, but have been living in Belarus for a long period. This self-inclusion of the non-ethnic Belarusians in the circle of the Belarusians together with the acceptance of this self-inclusion by ethnic Belarusians is a sign of social inclusion possible in Belarus. It demonstrates the type of inclusive resilience possible with regards to national self-identification.

Problems, challenges and the circle of support

In framing the emerging difficulties that FG participants face today, they mention economic problems. Among the problems, they mention the closure of some plants in the past. They also complain about specific challenges in the functioning of healthcare. Both youths and old-aged participants feel sorrow and pain that some young adults are inactive and passive, and do not use the opportunities available for them. Also, alcoholism among the population is mentioned.

An opinion is expressed that Belarusian tolerance supports people, assists them overcome problems, and helps stay calm: 'Tolerance allows you to be calm about such painful problems' (female, thirty-four, Brest). When solving their problems, people address their family, other relatives, and friends. The respondents mention that initially, they avoid addressing formal structures for help. Usually, they try to solve problems independently or address their folk. Further on, some do apply for professional aid in case informal structures do not manage to assist them: 'In my practice, I went to a psychologist working with child-related issues, and it was quite effective. So, you really shouldn't be afraid to address specialists and professionals who can help you as experts in the situation you are facing at the moment' (female, thirty-three, Brest). This opinion was supported by the other participants, demonstrating a clear vision of the sequence of actions in case one needs to ask for help. Facing a problem, you rely on yourself, address informal structures, and gradually apply to formal structures of professional assistance in case of trouble.

When dwelling on the relationship between the individual and the state, many respondents mention the dominance of a kind of paternalistic expectations in the Belarusian society: 'With us, well, me included, it seems to me that my comfortable life should be arranged by someone else. That is, I do not turn to, without even asking for help from the state, I expect that it should do something for me' (female, twenty-five, Minsk). The participation of citizens reduces to the role of the recipients of goods: 'We are used to blaming everything on the state, so we can say that our citizens have less initiative' (female, twenty-one, Mogilev). In turn, it leads to an unwillingness to participate in civil initiatives. Simultaneously, individual respondents are ready for organisation and express their desires to be socially active.

The issues of public safety are mentioned quite often. It should be noted that the discussion of this topic was initiated by the respondents themselves, which indicates its undoubted relevance. 'If you take the recent situation in Ukraine, I probably wish that children should not have to face such a situation. I mean the military conflict. I mean, naturally, security. Yes, and a peaceful life' (female, twenty-six, Minsk). Furthermore, the Ukrainian topic is not the only source influencing the perception of public safety. Often, even the experience of a short stay abroad (including the one in Western Europe and the USA) allows the respondents to tell that they experienced the situation when they felt insecure.

The starting point here was the Ukrainian armed conflict of 2014. It is necessary to mention that before this period when conducting similar studies, security issues did not cause a strong response among respondents (see in Bulynko, Danilov, and Rotman 2009, 10). Public safety and social peace were perceived as something permanent and taken for granted. Military conflicts in the world were presented as something extremely far away from Belarus, something that could not arise under any circumstances. However, the territorial and ethnic proximity of Ukraine influenced the public sentiment concerning security and peace in Belarus significantly.

Thus, the Belarusians see a clear sequence of actions they can take to address problems. They are ready to rely on themselves, address informal social structures, and apply to formal structures for assistance. The topic of public security is an issue relevant to the participants. They care about public peace.

Historical community of relations and community of relations today

A community can be a source of support for an individual in a troublesome situation. The ethnographic studies of the Belarusians describe a form of communal help, named 'talaka', which was present in the communal peasant life at the territory of Belarus and neighbouring lands till the middle of the twentieth century. According to the Encyclopedia of History of Belarus, 'talaka' is a folk practice of voluntary, free of charge mutual assistance in general labour activities to a community member or a family with the work, requiring a lot of workers and carriage, which one family could not cope with (like harvesting, scything, housing construction) and other general communal works done for the good of the whole village (channelling, well sinking, road making, etc.). Also 'talaka'—the invited neighbours, relatives, fellow-villagers—helped homeless fire victims, widows, wives of soldiers and those in real need (Pashkou 2001, 494). N. Karbalevich singles out the types of 'talaka': planned or regular, like harvesting, relatively irregular, for example, housing construction and emergency cases, such as those caused by fire, floods, etc. The author determines the following functions of 'talaka': communicative function, the function of socialisation, economic function, the function of maintaining social stability, psychological and ceremonial functions along with others (Karbalevich 2014). As the 'Belarusian Folklore: Encyclopaedia' states: 'Talaka' as a part of traditional lifestyle was represented in a special type of 'talaka' songs, verbal formulae of invitation to 'talaka' and gratitude for the communal help (Valodzina 2006, 592–593).

Thus, 'talaka' was realised at the level of horizontal connections. A community member knew that it was possible to ask neighbours or other members possessing certain skills, to invite 'talaka' for help. And the folk custom required people not to refuse the invitation to take part in 'talaka'. In return, any community member also participated in 'talaka' for the good of other families or the communal good. It was one of the forms of communal resilience, being a traditional, customary approach to solving the tasks that required communal voluntary work and mutual assistance.

Community-based structures can help people protect their interests. They exist today in other forms, though. The respondents mention their 'circle of contacts': their family, other relatives, friends they can apply to, asking for help in case of need. For instance, answering the question of who they can address in case of need when facing problems, they mention: 'Relatives. Close friends' (male, twenty-three, Grodno); 'I think so too, that we turn to family, relatives. They are the ones who will really help' (female, thirty-six, Mogilev); 'To relatives' (female, sixty-four, Gomel). Some are ready to take responsibility by themselves primarily: 'First of all, you have to decide for yourself and be ready to make decisions. But at the same time, if the situation requires it so that you would not become withdrawn. I think there should always be friends and relatives around you, first of all' (female, thirty-four, Brest).

According to the FGs data, neighbourhood communities, as well as other informal associations, do not play a significant role in the social life of the FG participants (except for single communities that have developed in the process of solving problems associated with shared equity construction, more on this further below). Nevertheless, fairly stable interpersonal relationships with individual neighbours are possible and mentioned by the respondents in the course of the focus groups: 'Yes. Neighbours at the summer cottage … occupy significant place in my life' (female, fifty-four, Minsk).

A relatively new channel assisting to 'reassemble' local communities is social networks in their broadest interpretation—from global options such as Facebook, Russian VKontakte or Odnoklassniki, to local Internet forums and chat rooms in messenger programs (like Viber, Telegram, etc.). As the focus group participants repeatedly note, residents of multi-apartment buildings create virtual groups in social networks or chats in messenger programmes: 'We have a group of our apartment building section in Viber' (female, thirty years, Minsk). This allows them to solve everyday problems quickly and to share relevant information. The starting point here is the need to solve a specific problem. At some stage of virtual interaction, participants often experience the need for face-to-face meetings offline. If the actions taken are successful, the cohesion of the group will increase, as well as the enrolment of new members into it.

It should be noted that the majority of urban residents live in apartment buildings in Belarus. The groups of participants in the shared construction of apartment buildings established via social media are examples of the functioning of such communities. Initially, such informal virtual communities are organised to exchange information between their participants on the progress of the construction. Consolidation of initially diffused and unstructured participants in virtual communities takes place in case problems associated with the construction appear. The members have to plan and implement specific actions aimed at solving their problems. At this stage of community functioning, as a rule, face-to-face communication of participants subjoins the virtual sphere of community-building. In addition, even when serious problems do not appear, boards of partnerships of owners (condominium associations) are formed from this kind of virtual communities after the commissioning of the house. The boards of partnerships start participating in solving their current issues. It means that initially spontaneously organized groups acquire official status. Even at the construction stage, when resolving the emerging issues, these groups gain experience while interacting with the official structures that have certain authorities from the developer to local government bodies.

The fundamental difference from the other network communities, in this case, is the fact that the participants in these groups know each other outside the virtual sphere. They inevitably meet face-to-face when establishing partnerships of owners, interacting in daily life as neighbours, as soon as they start residing in the building.

Another example of secondary consolidation via information technologies can be an organised interaction to solve business issues as described by one of the focus group participants. This person, who is an entrepreneur at a local market place, tells that he has created several regular informal chats with the other entrepreneurs from the marketplaces to solve work-related issues. Members of these groups exchange information with each other regularly (for example, about unscrupulous suppliers), which helps prevent or avoid problems. Gradually, communication in such groups may go beyond solving exclusively working issues. The participants can send birthday greetings to each other, congratulate each other on holidays, ask for advice or assist with resolving personal issues, or arrange joint vacations.

In all the above-mentioned situations, the information platforms serve as a powerful consolidating tool for creating virtual communities, which can unite the initially disengaged participants around the solution of a problem or task. Gradually, these groups can evolve and take other forms of community interaction. It is important to emphasize that the common task functions as an anchor point for the creation of virtual communities. While functioning, they can evolve into offline consolidated groups of neighbours or community members.

Visions of 'good life'

In considering their vision of 'good life', the majority of the participants mention two main components: moral satisfaction with their lifestyle (with the emphasis on work) and financial well-being: 'I would characterise such two aspects that are important for a person, for example, they should satisfy us.

Some moral aspect that is, if life brings you some moral, positive joy, impressions, and the second is the material aspect' (male, thirty-seven, Vitebsk). Notably, in all the focus-groups, harmony and satisfaction with one's life is put forward before material wealth.

Speaking about the intangible aspects of good life, the focus group participants often resort to such definitions as 'positive impressions', 'satisfaction', 'joy', 'self-realisation', 'I've found myself', 'to do my favourite thing', 'to do what I want'. Some FG participants define 'good life' via the well-being of their children: ' ... when the children are fine, we are fine then' (female, sixty-five, Grodno). It is the realisation of oneself in any professional occupation, the achievement of goals with the subsequent positive emotional impact that determines, according to the participants, their success and makes life good. The care of the older generation about the success and happiness of their children, which is also a life-long business for them, likewise fits into this context.

For the Belarusian participants of focus groups, financial well-being means a sufficient amount of material goods that are in direct accessibility. Despite the apparent naturalness of this situation, it is fundamentally different from what used to be observed in the late Soviet period. The availability of money in those days did not guarantee an easy acquisition of commodities (from automobiles to the most popular books). At the official level in the late USSR, there existed additional regulatory mechanisms for the distribution of consumer goods: such as queues, special distribution points and coupons. And that situation stimulated the development of informal social ties and formation of a kind of social capital—a circle of acquaintances who could help with purchasing some goods in short supply out of turn. Thus, the low efficiency of the state in resolving the issues gave rise to a compensatory mechanism on the part of the society. Certainly, it can be referred to as one of the forms of social resilience. However, in modern conditions with the abundance of goods and commodities available, the need for such informal social ties related to acquiring consumer goods has disappeared almost completely, which has contributed to the growth of social atomisation.

Another important point in the participants' vision of 'good life' is the availability of opportunities, possibilities for one's children and for oneself. For instance, a desire to have potential access and an opportunity to visit European countries was mentioned. Another respondent just states: 'For me, it is the availability of opportunities' (male, twenty-two, Grodno). It should be emphasised that here a hypothetical, abstract possibility is often described, and not one's specific plans. Thus, the availability of presumable opportunities and possibilities is an essential element of the vision of 'good life' for the Belarusian participators of the FGs.

It should be emphasized that, in the majority of cases, these are moral aspects to a good life: 'A good life is an opportunity for self-realisation and preservation of a certain subjective well-being' (male, fifty-five, Grodno). This indirectly indicates a sufficient degree of satisfaction of basic needs and the actualisation of those needs that belong to higher levels of hierarchy. Remaining within the framework of Maslow's theory of needs (Maslow 1987), one should note, that mentioning both the highest and the lowest levels of the pyramid of needs, the Belarusians thus strive to cover it as a whole: 'I would like to see health, well-being, prosperity and a decent life' (male, sixty-three,

Gomel). To a large extent, it explains the Belarusians' vision of 'good life' as a comprehensive idea and testifies to their awareness of its major sources and objectives.

The described perceptions of a community of relations and 'good life' are important for interpreting the Belarusian variant of resilience. They provide the nation with a framework and focus in its development.

The East–West axis

Resilience building at the international level is also a topic considered by the respondents. The visions expressed by the participants are the reproduction of the general public's interpretations of agenda-functioning in public and mass media discourses, which interact closely. To some extent, the general public's visions have to be taken into consideration by politicians and simultaneously reproduce the agenda set out by the national government.

A certain dualism is recorded concerning the framework of the East–West axis. As one participant states: 'Belarus has always floundered between East and West' (female, fifty-four, Minsk). The participants acknowledge a mix of quite a strong attraction to Russia, especially during the period when Belarus was a part of the USSR, along with their belonging to the Europeans, but with some stipulations: 'If you take an average cross-section of the country, I think we're somewhere between the Soviet Union, and then there's a Russian mentality and a more progressive western mentality' (female, twenty-six, Minsk). The respondents feel their unique features: 'It is believed that, yes - we are Europeans, but do not forget that we still have an irrational component, which is not inherent for the Europeans' (male, twenty-two, Grodno).

This dualism is also observed in participants' disagreements on the possible prospects for international cooperation and the guidelines that should be followed in further national development. A number of the participants believe in the western European model of development, noting its efficiency: 'The EU is necessary in any case' (female, thirty-six years, Mogilev). Other FG partakers, on the contrary, consider the western model of the world order to be unsuitable for Belarus. They often remark that Belarus would not enjoy the best role in such a case: 'I believe nobody is waiting for us in the west' (male, thirty-six, Minsk). The basis for such judgments is the observations on how the living standards and lifestyles of the Baltic countries, neighbouring Belarus (Latvia, Lithuania and Estonia) changed after their accession to the European Union: 'People used to go to Lithuania. It was so good. Some goods used to cost about twenty litai there, it used to be three euros, four. Now it costs twenty euros. And the Lithuanians cry' (male, thirty-six, Minsk).

Among the international organisations important for Belarus, the focus group participants mention the CIS, EurAsEC, EAEU, UN, the Union of Belarus and Russia, and the EU. In addition, the Eastern Partnership, CSTO and the Silk Road Economic Belt were named. The FG participants consider the CIS, the Union State of Belarus and Russia, the EurAsEC to be the most significant for their country. In mentioning that these organisations are important, the participants answer the question about the bodies Belarus can apply to in case of problems, and what enhances the country's international resilience in the event of external pressure.

The opinions about the Union State of Belarus and Russia differ. On the one hand, there are voices fully supporting it: 'The most important is the Union of Belarus and Russia and the economic community with Kazakhstan, the common customs space, the EurAsEC' (female, twenty-three, Grodno). Though, several participants mention that it is rather formal, inactive, supporting the CIS more: 'Well, they have already told that the Union State is more a formal organisation. And the CIS, it is very important' (female, sixty-five, Minsk). Notably, the attitudes to the EurAsEC are either neutral or positive; it does not receive negative comments at all: 'Well, I think we need, in principle, the Eurasian Union' (male, thirty-six, Minsk).

According to the FG participants, in the modern world, the state cannot exist without participation in international organisations, but this membership should be mutually beneficial: 'Therefore, alliances must be because in the modern world it is so international, one cannot survive. … Alliances must be, but they must be profitable, beneficial for all on some grounds' (male, thirty-five, Vitebsk).

Simultaneously, an opinion is voiced that there are so many integrative associations, and they duplicate each other and overlap: 'Now they are building the EurAsEC, but why do we need the CIS? The same thing after all. Why do we need ten different associations, if you can choose, for example, the EurAsEC, which is interesting for the same countries that are in the CIS? I also don't really understand politics, but I don't see the point in many integrative associations, which, in fact, are not clear' (female, thirty-six, Vitebsk). As one can see, the respondents support the participation of Belarus in political and economic alliances, see its participation as a useful form of a national strategy that enhances its strength and resilience, but they would like to see them as clearer structures, with simpler, not overlapping scopes and activities.

The FG participants consider, when it is necessary to protect the interests of the Belarusian economy, the country should rely on itself along with using the partnerships with other countries when it is mutually beneficial: 'Belarus should build partnerships solely based on its interests' (male, twenty-six, Brest); 'I believe that we should rely on our resources mostly, but also try to maintain peace with the allied states, try to maintain certain economic ties, partnerships' (male, sixty, Grodno); 'Simultaneously, for an average Belarusian it is more important to have relations with the closest states, our neighbours' (female, thirty-four, Brest). Hereby, the importance of collaboration with neighbouring countries is particularly emphasised by the participants along with finding other points for interaction and mutually beneficial cooperation.

The views of the FG participants on the Belarusian experience of resilience building in the international context, the experience of the country, positioned at the junction of two powerful geopolitical players, demonstrates the ideas that accommodate two counter-visions on the vectors of its possible national development. These competitive visions can contribute to the country's strength, enriching it via beneficial interactions with both the poles of attraction, on the one hand. On the other hand, these visions compete so that this competition weakens the interpretation of the prospects of development. Along with it, neither of the two vectors will solve the problems the country faces. Therefore, the parallel application of the ideas about the importance of the defence of the national interests, the concept of self-reliance, along with the support of mutually beneficial cooperation with the international partners and the neighbouring countries, can be a rational response, a balanced version of the Belarusian resilience in the international context. This simultaneous reliance on one's resources along with openness to cooperation within frameworks of international alliances and cooperation with neighbouring countries, when it is mutually beneficial, is a variant of the incorporation of the variety of visions in a complex but functioning structure. This multifaceted conception presents a Belarusian variant of inclusive resilience in the international context.

Conclusions

Premised on the unique first-hand data of the focus-groups and its analysis, this paper proves that the 'local' dimension matters for resilience building, including resilience building in the international context.

Belarusians' identity concepts, presented via the nominations of several typical referent groups, provide a set of values and attitudes that serve as an important part of personal and social resilience of the society. The Belarusians accept the representatives of other ethnic groups residing in the territory of Belarus. If these representatives express their aspirations to be a part of the Belarusian society and prove it by their actions—living long enough here and observing the way of life accepted in Belarus. This possibility for inclusion of the representatives of other ethnic groups is one of the manifestations of inclusive resilience.

Tolerance of the others who do not pose a direct threat emerges from the historical experience of the Belarusians. It is an aspect of the Belarusian strategy of responding to external influences. As a consequence, the Belarusian version of resilience implies avoidance of forceful solutions and open conflict scenarios when solving emerging problems.

Resilience formation at a micro-level involves those to whom a person can address in a difficult situation. Family is called by the participants in the first instance among those whom Belarusians turn to for help in a troublesome situation. Often it is only the family. Thereafter, it might be kinsfolk and friends from a prior student time.

Occasionally the participants mention virtual communities organised on territorial or professional grounds, though these virtual communities are not entirely virtual, because their members have face-to-face contacts finally. The existence of these connections is a source of social capital formation and constitutes meso-level resilience. In addition, the emergence of territorial and professional groups in hybrid space (virtual plus real) replaces the disappearing traditional horizontal connections and can be useful to people in solving problems that arise.

Perceptions of 'good life' and achieving some aspects of 'good life' is a part of the ability of the social system to choose the focus of development and to be resilient. The concept of 'good life' is presented in a complex and quite detailed way. It includes a set of elements repeatedly reproduced by several respondents. Most often it concerns both spiritual aspects, like self-actualisation and material components like financial well-being. People's aspirations are comprehensive. They simultaneously involve self-realisation and prosperity, are aimed at the sphere of the potential, expressed as the availability of opportunities. This complexity of perceptions makes it possible to compensate one at the expense of the other, to respond to a variety of types of challenges and is aimed at preserving development opportunities.

Among the social challenges, the FG participants repeatedly mention the fact, that a significant number of young people want to go abroad for permanent residence. This issue seems important to the FG participants because it may serve as a factor that weakens Belarusian society in the future.

The respondents support the international cooperation of their state, membership in alliances, the presence of allies and partners, interaction and cooperation with neighbouring countries. A division of opinions about the vector of cooperation exists: whether it is necessary to cooperate closely with western or eastern actors. However, the participants are clear that it is necessary to rely, first of all, on oneself when protecting the economic interests of one's country, being open to partnerships simultaneously.

The outcomes of the study (being based on the empirical data of 2019) have dual importance, and can serve as a pathway for further research in the context of the unprecedented social changes brought to the world, region and Belarus itself. The understanding of historically-conditioned inclusive resilience may be the key to developing a response strategy to contemporary challenges.

The obtained research results prove that resilience-building takes place not only at macro but also at micro and meso levels. The 'local' coping strategies can be traced, described and analysed even with reference to the international context. The Belarusians' visions of the problem coping strategies in the international context reproduce the scheme applied by the individuals at the micro-level: firstly, relying on oneself, secondly, addressing to friends and allies, further on cooperating with professionals. The emerging virtual or hybrid community groups contribute to resilience building at meso level, and strengthen horizontal social ties. The Belarusian variant of resilience is of relational, versatile, inclusive nature. The collected data add to the resilience theory, proving that the local coping strategies exist; they are relevant for the whole national system, even in the global context.

Disclosure statement

No potential conflict of interest was reported by the authors.

ORCID

Victor Pravdivets ⓘD http://orcid.org/0000-0002-1156-4413
Anna Markovich ⓘD http://orcid.org/0000-0001-5454-6103
Artsiom Nazaranka ⓘD http://orcid.org/0000-0002-7751-5665

References

Bourbeau, P. 2015. "Resilience and International Politics: Premises, Debates, Agenda." *International Studies Review* 17(3): 374–395.

Bulynko, D. M., A. N. Danilov, and D. G. Rotman, eds. 2009. *Cennostnyj mir sovremennogo cheloveka: Belarus' v proekte 'Issledovanie evropejskih cennostej* [Contemporary Person's Value World: Belarus in the Project 'European Values Study']. Minsk: BGU.

Chandler, D. 2014. "Beyond Neoliberalism: Resilience, the New Art of Governing Complexity." *Resilience: International Policies, Practices and Discour*ses 2(1): 47–63.

Chandler, D. 2020. "Security Through Societal Resilience: Contemporary Challenges in the Anthropocene." *Contemporary Security Policy* 41(2): 195–214.

Flockhart, T. 2020. "Is This the End? Resilience, Ontological Security, and the Crisis of the Liberal International Order. *Contemporary Security Poli*cy, 41(2): 215–240.

Hentschel, G. and J. P. Zeller. 2014. "Belarusians' Pronunciation: Belarusian or Russian? Evidence from Belarusian-Russian Mixed Speech." *Russian Linguistic*s 38: 229–55.

Karbalevich, N. M. 2014. "Talaka jak madjel' tradycyjnaga gramackaga dzenjannja belaruskih sjaljan (drugaja palova XIX – pachatak XX st.)" [Talaka as a model of the traditional civic action of the Belarusian peasants (second half of XIX – start of XX cent.]. *Vesnik BDU. Seryja 3, Gistoryja. Jekanomika. Prava [Journal of the Belarusian State University. History. Economics. Law]* 3(1): 14–17.

Korosteleva, E. A. 2020. "Paradigmatic or Critical? Resilience as a New Turn in EU Governance for the Neighbourhood." *Journal of International Relations and Development* 23: 682–700.

Korosteleva, E. A. and T. Flockhart. 2020. "Resilience in EU and International Institutions." *Contemporary Security Policy* 41 (2): 615382–700.

Maslow, A. H. 1987. *Motivation and personality*, 3rd ed. Delhi, India: Pearson Education.

Pashkou, G. P. 2001 ch ed. *Jencyklapedyja gistoryi Belarusi* [Encyclopedia of the history of Belarus] in 6 vol. Vol. 6. Book. 1: Pusyny—Usaja. Minsk: BelEN.

Pivavar, K. S. 2015. *Belaruskaja mental'nasc' u mounaj prastory mastackaga* tjekstu [The Belarusian Mentality in the Linguistic Space of the Literary Text]. Vitebsk: Vitebsk State University named after PM Masherov.

Samusevich, V. M. 2012. *Belaruskija SMI u lagasfery nacyjanal'naj kul'tury* [The Belarusian Mass Media in the Loso-Sphere of the National Culture]. Minsk: BDU.

Statistical yearbook. 2020. Minsk: National Statistical Committee of the Republic of Belarus. https://belstat.gov.by/upload/iblock/7d0/7d0ed3586722991264205df8d056cf60.pdf

Valodzina, T. V. 2006. Talaka [Talaka]. Belaruski fal'klor: Jencyklapedyja [Belarusian folklore: Encyclopaedia] in 2 vol. Vol 2. Ed. GP Pashkou. Minsk: BelEN.

The Azerbaijani resilient society: explaining the multifaceted aspects of people's social solidarity

Azer Babayev

Kavus Abushov

Abstract   *This article addresses both a conceptual and empirical gap in research by examining social identity and resilience in Azerbaijan. Based on unique first-hand focus group data (FGD), it discusses Azerbaijani society and its resilience-building with a focus on its temporal dimension—the past, present and future, while the present here is to be considered as a nexus thereof—especially in terms of the idea of 'here and now that matters'. In this way, it refers to resilient society, premised on past memories and traditions, as well as existing support infrastructures (e.g., family, kinship, neighbourhood). It is further bound by affective solidarity as a coping mechanism of local communities to withstand adversity and even war in Azerbaijan, and to grow into 'peoplehood' when fully mobilised.*

Introduction

The emergent Complex International Relations theory (CIR) (Kavalski 2007, 2016; Korosteleva and Flockhart 2020; Korosteleva and Petrova 2020) has put forward several important assumptions—including non-linearity, self-organisation ('emergence'), society as a complex system, and relationality (see the Introduction to this Special Issue)—that naturally challenge the effectiveness and, for that matter, the entire possibility of the top-down international governance paradigm.

Global problems, CIR asserts, can be most effectively and sustainably solved at the local level by building on indigenous factors such as existing social structures, traditions, philosophies, and aspirations, instead of importing and imposing one-size-fits-all solutions. The CIR insights have developed alongside the post-colonial and post-structuralist approaches, which concur the importance of 'the local' and argue for pluralisation of the study of International Relations by paying particular attention to non-Western social

Research for this article has been carried out within the GCRF COMPASS project (GCRF UKRI ES/P010849/1, 2017–2021) conducted by University of Kent and Cambridge University, and supported by UK Research and Innovation (UKRI). The authors thank Elena Korosteleva, Irina Petrova and other colleagues of the GCRF COMPASS project for their valuable feedbacks as well as the ADA University for the additional support given to them during the research. They also wish to thank the journal's editors and anonymous reviewers for their useful comments on the earlier versions of this article, which greatly helped to improve it.

structures. As mentioned in the Introduction (Korosteleva and Petrova 2021; see also Bousquet and Geyer 2011; Nordin et al. 2019), social orders emerging as a result of non-linear development and by passing through a number of feedback loops facilitate the resilience of a social system. Consequently, each system (society; community) aims to maintain its self-reliance and self-governance by being able to draw on its unique inner strengths and capacities, and to transform in the face of adversity.

In line with these theoretical considerations, this article aims to examine the importance of the 'local' through an empirical analysis of the case of Azerbaijan. It asks a simple, yet important question of *what local structures and practices make Azerbaijani society resilient to face the challenges of time, and to transform in response to complexity*. In answering this question, the article illustratively sheds light on (a) what resilience may look like at the community level, and particularly focuses on (b) how the Azerbaijani people's identification as a *process of becoming* (Chandler 2021; Korosteleva and Petrova in this volume), driven by 'the good life' visions (ambitions for better future) and community support with its local coping strategies, contribute to the resilience of the Azerbaijani society. In line with Berenskoetter (2011, 652), the 'becoming' approach here serves as a useful way to think about identity (formation) in terms of being 'manifested through the future' and its visions.

Contrary to the widespread approach to the study of resilience through the analysis of political regimes and state institutions (Rouet and Pascariu 2019), the research purpose here is to explore *people*'s social solidarity and the making of resilient communities in the country, putting *societal* resilience at the heart of the analysis. By doing so, the article aims to contribute to three bodies of literature. First, as mentioned above, most of the literature on resilience has focused on state institutions.[1] Analysis of *societal* structures contributes to the more complete and nuanced understanding of resilience. Second, the study adds to filling the knowledge gap about *community resilience* by empirically unpacking its elements, as theorised by the Introduction (Korosteleva and Petrova 2021). Third, answering the question of what makes Azerbaijani society resilient is of utmost importance for the revision of international governance, based on a more local-sensitive approach, thus addressing the plea to 'decentre' the study of international affairs (Fischer Onar and Nicolaïdis 2013).

Situated in Central Eurasia and ultimately at a crossroads of the East and the West, the Global North and the Global South, Azerbaijan presents an important case to study societal resilience. As such, it is part of one of those world regions where there are, to varying degrees, several global problems such as wars, refugees, human rights, democratisation as well as poverty and environmental issues at the same time. Also, the troubled past of the country including radical sociopolitical transformations in the last two centuries as well as its enduring rivalry with the neighboring Armenia and the ongoing tensions around its Nagorno-Karabakh region constitute a unique 'complication' to test

---

[1] In the same way, often focused on the illiberal or authoritarian practices of the political system, work on the post-independence development of Azerbaijani society largely remains macro-level- or state-centric and, even when making micro-level claims mostly do so without an explicit analytical framework (see for example, Cornell 2011; Hirose 2016; Ergun 2021).

societal coping. Most crucially, the resilience-building capacity of people's self-identification and a sense of community—especially in local settings—deserve serious academic consideration, especially when addressing fragile communities in conflict-ridden geographies.[2]

Following the framing of this Special Issue, the article proceeds by providing an analytical framework for the study of *societal resilience* in Azerbaijan. In a common academic and political understanding, resilience is often perceived as an adaptive 'quality of a complex system' that enables entities to respond more adequately to change and to make them enduring. In this view, resilient communities are those that are capable of withstanding, 'bouncing back' to normal and adapting in the face of challenges (Bourbeau 2018; Rouet and Pascariu 2019; European Commission 2020; Cusumano and Hofmaier 2020; Tocci 2020). To move beyond this approach to resilience, which rather simplistically implies a linear development, we prefer a more comprehensive understanding of the concept in a world of complexity. As developed in the Introduction to this Special Issue (Korosteleva and Petrova 2021), resilience is understood here also as an 'analytic of governance' both in terms of thinking and (self-)governing, which builds upon local definitions of potential challenges and draws on self-reliance, and self-organisation that mobilises communities' own resources and capacities in the face of adversity. Accordingly, to understand what makes communities adapt and endure, and how they are to be governed today, the concept of resilience is explored as an overriding analytical framework, with its constitutive elements—identity, the 'good life', local coping strategies and support infrastructures, which, when mobilised, turn communities into 'peoplehood' in the face of adversity (for a further discussion of the concept, see Korosteleva 2020; Korosteleva and Flockhart 2020; Korosteleva and Petrova 2020).

Based on this analytical framing, resilience (-building) in Azerbaijan is operationalised and illustrated empirically in terms of its multiple key components in practice such as (a) people's identity and the 'good life' aspirations, (b) local support infrastructures and (c) the mobilisation potential of peoplehood, particularly with regard to how these elements combine to represent and enhance societal resilience. In this context, the article discusses as key local coping mechanisms, the community solidarity and its tangible incarnations, with a salient *affective* element, building upon Durkheim's notion of social solidarity.[3]

Azerbaijani identity as a process of becoming

To develop a better understanding of social resilience in Azerbaijan, the article first considers Azerbaijani identity, as a process of becoming, to be made complete and thus meaningful by a sense of a 'good life' as 'rational dreaming'. Notably, it is people's identification that sets the stage for community

---

[2] In the last three decades, several outbreaks of communal violence, or even war in the region (including a second all-out Karabakh war with Armenia in 2020) show how adverse the broader context for local communities is.

[3] Emphasising the non-political, non-economic foundations of society/social order, Durkheim (1984) defines *social solidarity* as a set of (cultural) bonds that hold people together and unite them into a society—a social whole with a 'collective consciousness/conscience'.

solidarity and the relevant support infrastructures, which in turn opens the possibility for a (capable/powerful) peoplehood, or the moment of resilience (Korosteleva and Petrova, in this volume). This article illustrates how these components work together to contribute to making the Azerbaijani society more resilient in addressing the relevant (and potentially global) challenges.

The Azerbaijani identity can be conceptually characterised through a combination of 'hold' and 'pull' factors, which in turn help to enhance the social resilience of Azerbaijani people on both national and local levels—which we term here as *resilient identity*. A useful way to treat this kind of identity is in terms of defining it not only by one's past (experiences) but also by their future (visions). It is especially the future dimension of people's self-identification that makes it a resilient kind of identity. As Nietzsche (1889, 2) puts it, he who has a why (i.e., purpose/future vision) to live for, can bear almost any how (i.e., current adversities/challenges).

The 'hold' factors refer to the material/experiential (i.e., experienced) aspects of *resilient identity*. As such, they imply a process of staying stable and anchoring through the past (and present). The 'push' factors, on the other hand, describe immaterial (ideational and thus yet unexperienced) sources of identification, which mainly refer to the future, and to the 'rational dreaming' about a 'good life'. Indeed, future is always an *idea*, and this, as part of people's *existing* identity: that is, thinking and planning of future amounts, then, to the materialisation (realisation) of one's current ideas and visions. Hence, at the end of the day, a person is neither their past nor their present; they are their future (cf. Berenskoetter 2011). In other words, nothing has really happened until everything (i.e., prospective future) has happened.

In Azerbaijan, a distinctive duality of people's identification is conspicuous: divergence of identification at macro/national and at micro/local levels. At macro level, Azerbaijani identity 'has historically been fed by diverse cultural wellsprings: Turkic nomadic epics and language; Shia Islam from Persia; Sunni Islam originating with the Arabs and taken up by the Ottomans; administrative culture and industrialisation from Russia; and Western values from Europe' (ICG (International Crisis Group) 2004, 8). As for the sources of identity construction in Azerbaijan in a (modern) historical perspective, Azerbaijani identity was moulded in different political regimes (Russian Empire, brief 1918–1920 independence, Soviet Union, post-Soviet period) with their own pull and push factors. Overall, considering its recent dynamics—notably that throughout the twentieth century—a transformative process in people's own national identification has prevailed, moving away from a narrow (ethnically defined) concept of Turkism towards Azerbaijanism—a broader civic notion of national identity without rejecting the Turkic element of Azerbaijani identity (Babajew 2010).[4]

Although Azerbaijani people are united generally around a shared understanding of history as well as common culture including values and norms, the micro/local process of self-identification has historically oscillated between two extremes: religious and communal identities. On the one hand, people saw themselves as Muslim, but on the other hand as members of their

---

[4] For a detailed account of Azerbaijani identity including its historical roots and evolution, see Altstadt 1992; Shaffer 2002; Moreno 2005.

respective clans or tribes. In the tsarist period, the (Turkic-speaking) population of the Eastern Southern Caucasus were, for example, simply referred to as
'Muslims' (besides 'Tatars/Turks'), both by themselves and others (Baberowski
2003, 26; Motika 2009, 300). For this very reason, some Western authors only
moderately assume the existence of 'Azerbaijanis as a homogeneous ethnic
group' in the pre-national period (Kappeler 2001, 142).[5] Overall, identity formation in Azerbaijan has historically been highly dynamic, which this article
aims to take stoke of as an essential element of societal resilience.

Methodologically, the article draws upon unique data derived from focus
groups conducted in the major cities of Azerbaijan, namely Baku, Ganja, Shaki,
and Lankaran. The rationale behind concentration upon the major urban communities is that they encompassed a larger mass of people with diverse backgrounds, and in the case of Lankaran and Shaki in particular, they also
included people who did not directly reside in these cities but around them.
Thus, whereas one can attribute life in the capital Baku to urban *strictu sensu*,
one cannot do the same for Shaki, Lankaran and Ganja (to a lesser extent). In
the latter, the distinction between urban/rural is not strict and at times
blurred. Such a mixture is believed to have provided representativeness to a
large extent. The empirical data derived from 10 focus group discussions (four
in Baku, two in each of the three other cities) with an average of 15 participants from different age groups, conducted in the course of 2019. For the purpose of diversity, attention was paid to the composition of the focus groups, it
included people from different age categories, as well as social status, e.g., students, civil servants, journalists, teachers, doctors, activists and pensioners. The
focus groups were conducted by a moderator (an author of this article), an
assistant and an observer. Each focus group lasted for 90–120 minutes. Many
of the issues discussed in the current paper derive their data directly from the
focus groups; these included questions of national self-identification, identity
and good life aspirations, understanding of community and resilience, external
perceptions and various aspects of social relations.

The FGD was analysed using constant comparative method and classical
content analysis. The data from the multiple groups was used to assess if the
themes that emerged from one group would also emerge in the other groups.
This served the purpose of allowing to reach data saturation. Thematic patterns in responses were categorised into groups and coded accordingly. Any
serious within-group dissent was also incorporated into the data. The analysed
data were derived from the text of the focus groups that was transcribed and
translated into English.

Most importantly, the data collected through the focus groups served to
answer the central research question of what makes Azerbaijani society more
resilient. In particular, the focus groups helped to answer in-depth the questions of which practices at the local level have enhanced the resilience of the
Azerbaijani society, and how unique the local workings of resilience in
Azerbaijan are.

---

[5] Thus, as in line with the Islamic teaching that the primary identity should be Muslim and
that of 'ummah' (rather than the nation), Islamic identity had prevailed at the time, postponing
the need to develop an ethnic identity until early twentieth century.

Some of the focus group questions were open-ended questions, whereas others had specific numbers to measure their impact. For example, the question: how satisfied are you with your life when compared to living standards in other similar countries, or: what is the level of trust and security, were directly measurable, whereas questions such as: what kind of future do you see for yourselves (a question that linked identity with the future) and your families, required certain interpretation. These questions helped to understand how resilience (as self-governance/-organisation) works in practice in Azerbaijan, how the hardships the country has lived through since independence have shaped its level of solidarity, and how the global challenges are experienced and dealt with by people on the ground. In particular, it has sought to understand how relevant support infrastructures of communal life such as family, kinship and neighbourhood have contributed to resilience in Azerbaijan. And how this process has been driven by a specific (i.e., affective) form of solidarity.

It should also be noted that the research, being a small study dedicated to explaining the societal resilience in Azerbaijan, has had limitations. It is primarily focused upon the Azerbaijani society on the whole. For the reasons explained above, the focus groups were held in the major cities, although much effort was paid to make them as representative as possible, especially in terms of covering all four geographic areas of the country (Baku in the east, Lankaran in the south, Ganja in the west, and Shaki in the north of the country). Therefore, when making generalised claims in the article, the authors are also aware of the limitations of their arguments.

Azerbaijani people and a sense of community

How does people's identity (identification and perceptions of good life) and solidarity contribute to the resilience of the Azerbaijani society? Drawing on primary FGD (Focus Group Data) (2019), this section discusses Azerbaijani people's identity as leading to communal solidarity in terms of local relations and practices in the country.

As the data confirms, at the micro level, there is a high sense of community, or oneness, among people in Azerbaijan. In other words, a strong 'we' feeling exists among community members: that is, a consciousness of kind or an awareness of close association—and a consciousness assumed by Azerbaijani people by virtue of their membership in the community.[6] Specifically, this collective consciousness is structurally twofold and composed of a shared past—a *bond of fate*, and a commonly desired future (a sense of 'good life')—a *bond of destiny*; and the lived present as their nexus links both of them, and thus makes the whole a reality of people's daily life. As such, it shapes Azerbaijani people's identity, the way they perceive the world around them and how they act on their perceptions. It is a worldview whose core values are people's survival and thriving in their local environments often

---

[6] As it was mentioned in one of the focus groups in Baku (FGD 2019: male participant, 34 years old, civil servant), one's membership of community may be shaped by different factors, such as existence of common problems, seeking of a common future, common past etc.

affected by global challenges and crises, as well as their contribution to a good life.

Historically, already in the nineteenth century (in the context of Russian colonialism), community life and local relations in Azerbaijan showed significantly more modern structural characteristics than, for example, the tribal communities in Central Asia (Auch 2004). Later, throughout the twentieth century, Azerbaijanis' identification evolved along the civic lines, and the dominant form of identity in all layers of the society was Azerbaijani.

Generally speaking, the historical memory is an important source of Azerbaijani identity, especially related to the relatively recent experiences of the nation.[7] This firstly concerns the Stalin repressions of the key intellectuals of the country in 1930s and the deep trace it left within the historical-intellectual memory (Baberowski 2003; Cornell 2011). During the Soviet Union, this memory was well embedded into institutions, such as school and university curricula and public debates in 1970–1980s. Equally, Azerbaijan's participation in the World War II and the large number of losses as well as famine had also left a trace in its historical memory, although this had played a lesser role in the shaping of the national community than the repressions (Cornell 2002). Another important reference point in the historical memory, that became prominent in the early years of independence in particular, was the Azerbaijan Democratic Republic (ADR, 1918–1920), which, with its democratic traditions, has acted as a 'source of pride and point of unification' (FGD 2019: 41-year-old male participant, school teacher of history in Lankaran).

In the recent decades, people's sense of community has also considerably strengthened. The events and processes that have contributed to this include the freedom movement of late 1980s that led the process of gaining independence from the Soviet Union, state- and nation-building policies as well as the war with Armenia. The independence movement is especially important in this respect, and it was some form of an outburst of an affective solidarity of the nation. The movement had started around the Nagorno-Karabakh dispute, namely in response to the demands from Armenia and the Armenian community of Nagorno-Karabakh for the unification of the two in 1988.[8] People's response to a violent crackdown by the Soviet leadership on civilians in Baku in January 1990 (called the Black January), demonstrated a high level of social solidarity, when people, independent of their ethnic identity, gathered around the central idea of getting independence from the Soviet Union and unifying around a sense of community. The Black January as well as the subsequent tragedies during the Karabakh war, especially the Khojaly massacre in 1992 has reinforced the sense of community further.[9] Thus, overall, a sense of community in Azerbaijan derives from the common culture, values, historical memory as well as the thorny path to independence (Croissant 1998; de Waal 2003).

---

[7] In almost all the focus groups, emphasis was put on the role of historical memory in the identity formation.

[8] For Azerbaijani independence movement, see, for example, Motika (2009) and Cornell (2011).

[9] In the night of 25–26 February 1992, in Khojaly, a village in the Nagorno-Karabakh region populated by Azeris, Armenian troops killed more than 600 civilians, including women and children—making it the largest massacre in the entire post-Soviet space. For a detailed account, see, for example, Human Rights Watch/Helsinki (1994); Goltz (2015).

Today, Azerbaijani identity's (aka Azerbaijanism's) *adaptability* to, and *blending* with, its cultural and historical neighborhood in the broader region is remarkable—especially in terms of representing people's resilience-building. For example, in terms of religion, Shia Iran constitutes an important reference point, while, linguistically, Turkey does the same for people's self-identification.[10] And their daily practices and customs are to a large extent shaped by a Caucasian and Russian way of life. In doing so, Azerbaijanism appears to be more inclusive than exclusive in socio-cultural terms. Its core is the vision of Azerbaijani people as a cultural community/nation linked by the existence of a shared/common historical and geographical entity. As the poet John Donne (n.d., 135) famously put it, 'no man is an Island entire of itself; every man is a piece of the Continent, a part of the main'. This idea applies also to the importance of the notion of 'being part of a broader community/-ies' in Azerbaijan, and it highlights the degree to which Azerbaijani identity is fashioned by historical interaction with, and membership of, various regional traditions and cultural worlds.

That is why Azerbaijanism today connects easily to different, even opposite *worlds of identity* and peoplehood in the broader region: Turkish, Muslim, Caucasian, Iranian, Russian and/or European. Such an attitude to identity was observed in most of the focus group discussions (FGD 2019). Being lived at the micro level, this adaptable and inclusive identity of people amounts to living in Azerbaijani *hamsoya*—as a shadow of each other, and can even be at odds with nationalism at the macro level (see Nurulla-Khodzhaeva; Korosteleva and Petrova in this volume). For example, this factor explains, at least in part, a high number of ethnic Armenians living in Azerbaijan and their social acceptance at the communal level.[11] Thus, what is incompatible becomes compatible within the 'local' context of Azerbaijanism. This is also one of the key reasons why ordinary Azerbaijanis, who migrate to neighbouring countries such as Turkey or Russia, often display a high degree of substantial integration to those societies, unconstrained by their ethnicity. This point was reinforced by most focus group participants when asked if and how they could adapt to living outside Azerbaijan (FGD 2019).

Yet, what makes Azerbaijani people's inclusive self-identification socially visible, tangible and thus resilient? When considering the social identity (in the form of a group cohesion, community, or sense of belonging), the article refers here to *solidarity* as a key concept that describes what holds people together and contributes to their resilient identity. In this way, it constitutes the working mechanism of identity, which implies one's *active* identification as a part of their community—i.e., some kind of investment in their group or community (see, for example, Melucci 1996; Scholz 2008; de Beer 2009; Cureton 2012). Also, being 'intrinsically relational', solidarity implies in a rational way not only people's recognition of their mutual interdependence, but also the acceptance of common responsibility between them for some state of affairs and a response entailing reciprocity, not exchange, coercion, or strategic concession of one to the other (Archer 2013, 5).

---

[10] Also politically, Turkey is seen as a key reference. In all the focus groups, it was unanimously emphasised that the only external actor that Azerbaijanis fully trust is Turkey, which is seen as a fraternal nation (FGD 2019).

[11] According to various estimates, the number of Armenians living in Azerbaijan (outside Nagorno-Karabakh) is around 20,000 to 30,000, and they mainly comprise persons married to Azerbaijanis or of mixed Armenian-Azerbaijani descent.

Perhaps more importantly, solidarity can also be referred to as a feeling or as involving relationships characterised by an emotional bond (for example, see Hunt and Benford 2004; Scholz 2008; de Beer 2009)—highlighting its affective dimension. It is the emotional identification that, in fact, allows people to see themselves in terms of belonging to the group and to shape their identities around membership within the community (Melucci 1988; Rorty 1989; Hunt and Benford 2004; Hooker 2009). This very dimension of social solidarity explains its key importance as an underlying factor for communal resilience-building in Azerbaijan because it stands out as being a crucial coping mechanism for the people and their communities.

Identity and a good life's aspirations

A useful way to think about the community—'the local' in Azerbaijan—and to understand it, is through the people's identification and aspirations for a 'good life' to begin with. At the micro level, they are the first important factor for keeping a sense of community and becoming, in a sense, an *ideational* means of 'escaping forward' from reality's *material* complexity.

Should we assume the primacy of the local and the indigenous, the simple yet important question here is: What are the existing local structures and practices that make Azerbaijani people and their community resilient? Most importantly, we need to understand *the local way of being and becoming* in the country, inclusive of its social and cultural adaptability to the challenges of an increasingly complex world.

Firstly, how do people and communities perceive their identity and their future in Azerbaijan? What constitutes the existing we-feelings? At the macro level, most people today see themselves culturally as 'proud' Azerbaijanis (Caucasus Barometer 2013; FGD 2019, Baku, Lankaran and Ganja).[12] Our focus here is, however, primarily on the micro level of Azerbaijani identity as a key source of people's resilience-building.

Still being a pronounced form of social identity, *localism* and by extension *regionalism* are widespread in Azerbaijan and reproduce themselves across the country including the capital of Baku. Overall, this pattern roughly corresponds to the borders of the former Azerbaijani *khanates* such as Baku, Karabakh, Erivan, Nakhchivan or Ganja.[13]

Although the primary loyalty of Azerbaijanis does not lie with their respective regional groups thus remaining subordinate to the macro-level of overall national identity, many still tend to 'appreciate' their regional identities alongside—such as Karabakhi, or Nakhchivani, for example.[14] This can even occur when making new friends or acquaintances in daily life. However, as

---

[12] This idea was unanimously shared by the participants of the focus groups in all different cities of Azerbaijan. It was mentioned that a strong all-encompassing idea of Azerbaijani meant a lot to the people, because the concept is more than a civic/ethnic definition only, more of a unifying concept that includes the difficult historical path.

[13] Being ruled by a monarch or military leader, the khanate was a historical state formation in Eurasia—just like a medieval feudal state in Europe.

[14] In all the focus groups, it was unambiguously emphasised that an all-encompassing Azerbaijani identity matters most, only secondly comes the regional identity. In some of the groups (e.g., with the 50+ generation in Ganja), it was emphasised that Turkic identity should be emphasised alongside with the Azerbaijani one.

several focus group participants in Baku shared, the pattern of regional belongingness may gradually be submerging and become marginalised in the last few years due to economic development and modernisation. As one participant put it concisely, 'in the light of the changes, old allegiances and customs have given way to newer and other forms of identification' (FGD 2019: 54-year-old female participant, school teacher in Baku). Thus, one of the key questions asked about one's social background is about their regional origin: not only the answer to the question *'where do you come from?'*, but rather *'where do you originally come from?'* is therefore essential for a deeper 'social communication' (Deutsch 1972) among Azerbaijanis.[15] As such, regionalism provides people with a normative basis for their social action. Indeed, local, or regional identities still seem to be important in Azerbaijan, which projects itself even onto domestic politics.[16] This is the case even though almost all members of the country's elite are officially committed to the Azerbaijani nation.

Besides regional identities, *religion* constitutes another important source of Azerbaijani identity at the local level.[17] In addition to being an important part of people's national identity and thus providing them with a collective sense of belonging (on the macro level), religion locally provides a powerful ideological mechanism of survival and adaptation for the individual in a complex and uncertain social environment. As one focus group participant succinctly stated, 'religion is a last resort in hard times, when all other remedies have been exhausted' (FGD 2019: 36-year-old female participant, employee at a cultural centre in Shaki).

In general, people see the role of religion as important in their daily lives and identity, even though most of them are not practicing believers (Caucasus Barometer 2013; FGD 2019). In Azerbaijan as a Muslim society, Islamic values and beliefs act in two ways as a source of identity for the individual. First, they help attain personal fulfilment, and empowerment; it is not surprising that the Islamic phrase *InshAllah* (God willing) is one of the most widely used expressions of encouragement in everyday life when speaking of future events that one wishes or hopes will happen. Second, Islam encourages to help the disadvantaged members of community.[18]

Most notably, religion has been one of major ideological alternatives to fill the relevant vacuum after the collapse of communism. For many, Islam has provided hope in the midst of despair, whether that hopelessness was due to personal/social issues or to economic deprivation such as unemployment and loss of income following the end of the Soviet Union (Najafizadeh 2012, 93–94)—and it still continues to provide not only a worldview, but also a

---

[15] It implies that one is also interested to know where your (grand-)parents come from.

[16] For more on informal practices in Azerbaijani politics, see, for example, Altstadt 2017; Safiyev 2018.

[17] We examine here the general role of religion as a key social/cultural force without going into particular specifics, though it should be noted that the level of religiosity varies in different parts of Azerbaijan, and there is a sectarian divide (i.e., Sunni/ Shia). For the dynamics of religion in the post-independence period, see, for example, Ismayilov 2018.

[18] As one of the five pillars of Islam, *zakat* (to donate a certain portion of one's income for charitable causes) ranks very high in importance in Muslim societies. Accordingly, the general idea of donating/supporting people in need is also a widely internalised social value across local communities in Azerbaijan; such popular moral sayings/codes as *'Kasıba əl tutarlar'* (the poor need to be helped) alone speak for the prevalence of this social disposition.

vision, a path, guidance and a direction in life at a personal and/or community level (Ismayilov 2019).

In addition to collective memories, norms, and solidary affects, collectively held visions and/or *shared aspirations* also act as a driving force of collective identity and, by extension, of resilience-building. That is especially because historically Azerbaijani national identity is relatively young and thus, nation-building as a historic process is not yet entirely completed, especially in its political dimension. Thus, it is still 'mission unaccomplished', a process where the vision of the future matters today.

In the context of Azerbaijani identity today, what are the communal visions, aspirations, and images of the future? As FGD (2019) broadly suggests, a *good life*'s aspirations are certainly an important part and driver of people's identity, which manifest themselves at both national and local levels of societal life, and as such, hold people together and make them resilient. They are essential in creating a sense of community that binds individuals across society into a social whole with a powerful collective consciousness and a continuing shared culture.

Here, related to global challenges such as peace/conflict resolution and democratisation, two most important examples for a *good life*'s aspirations stand out at the macro level of Azerbaijani identity, which in turn appears to be still incomplete from this perspective. Accordingly, the realisation of these two goals could seemingly make an Azerbaijani nation-building process (politically) complete: first, the overwhelming majority of people have a strong aspiration to see the Nagorno-Karabakh conflict solved sooner rather than later and thus the territorial integrity of the nation restored.[19] In general, restoration of the territorial integrity can be seen as the most significant ('aspirational') element of the national identity. In this regard, there is an intimate relationship between these two elements. In people's perception (as it was shared by most focus group participants in Baku), it represents the physical aspect of Azerbaijan's identity, the very *ground* of its existence and its historical continuity. As Romano (1947, 56) puts it, the state does not so much have a territory, rather it is a territory. In that sense, Azerbaijan does not so much have Karabakh, rather it is the Karabakh that matters as a dominant sentiment of people. In this way, this aspiration can be even viewed as the biggest collective 'Azerbaijani Dream'—consequently, many are willing to achieve it even through personal sacrifice, or enormous risk-taking, as it was demonstrated by the 2020 Karabakh war. As such, it plays an essential role in empowering the Azerbaijani society as a whole, and especially creating a unique sense of social cohesion and stability that enables the individual to engage in cooperative activity with other members of the nation.

Secondly, given that Azerbaijan represents a strongman rule with highly centralised government power, building a democratic country is another

---

[19] It is not surprising that in public opinion the Nagorno-Karabakh conflict ranks first among the most important issues facing the country (e.g., before unemployment). Also, overwhelming majority of people claim that they would never accept Nagorno-Karabakh as an independent state (81%) or as a formal part of Armenia (96%) (Caucasus Barometer 2013). The conflict was also unanimously presented as the number one challenge of the country in all our focus groups (FGD 2019): concepts like good life aspirations, building a good future for the next generations were linked to the restoration of Azerbaijan's sovereignty over Karabakh.

equally important national aspiration for at least significant segments of society.[20] The post-World War I ADR too serves as the most important ideal for *becoming* a democracy.[21] Thus, while solving the Nagorno-Karabakh conflict could make the *territorial integrity* (and a connected to it identity) of the Azerbaijani nation complete, building democracy would do the same with its *political* identity.[22] This point was well-reflected upon in the focus groups, and as one focus group participant passionately put it, 'there is a need for an active civil society despite all the political challenges, and the only real way of building it, is acting from below' (FGD 2019: 23-year-old male participant, student in Baku). They were unanimous in the need to democratise and develop a strong civil society (since the two are interrelated).[23]

Just as state independence in 1991 brought external freedom for the country (i.e., national freedom), so a would-be democracy could lead to the internal freedom of people (i.e., individual freedom). This collective aspiration also creates social solidarity and serves to integrate the behaviour of individuals into cooperative activity with other like-minded members of the society, or certainly has the strong potential to do so.

These two aspirations can be characterised as a combination of both 'robust' and 'creative' visions (Berenskoetter 2011) that are driven by those ideas that not only connect the Azerbaijani past (e.g., ADR) with a desired future, but also create a sense of becoming, which can radically transform today's political *status-quo*. Also, these visions can both be considered *attractive* for sharing among diverse groups and people in the country, because they resonate with the nation's 'glorious and heroic' culture/history, and *possible* to realise, due to the mobilisation potential premised on the knowledge of the nation's available resources and resolve.

At the micro level, too, *good life* aspirations are a salient aspect of the Azerbaijani identity and operate as an important social force of people's identification. These again create interpersonal unity and serve to integrate the behaviour of individuals into local and communal enterprises. Here most importantly, a rather collectivist than individualist sense of 'good life' helps facilitate people's and/or communities' social cohesion—we-feelings in immediate social settings. In this regard, historically two totally different legacies are coming uniquely together—the collectivist legacy of communism and that of Islam.[24] À la 'happiness shared is a double happiness' (as one female focus group participant noted in her definition of happiness), this attitude is

---

[20] According to Caucasus Barometer (2013), overwhelming majority of Azerbaijanis show positive attitude towards democracy.

[21] In all the focus groups, the ADR was referred to as a key source of national pride (FGD 2019).

[22] Particularly in the focus groups in Baku, it was emphasised that a strong society/state is the one with an organised, dynamic civil society and eventual democratisation.

[23] Especially in the focus groups in Baku, a large majority of the participants emphasised the need to develop a stronger civil society, a stronger link between the society and the state. And this would be an essential part of people's good life aspirations.

[24] It is not surprising that in the early Soviet period the Azerbaijani leadership believed even in a congruence of Islam and communism. The first Soviet leader Narimanov made a case for it: you can convince our mullahs by making references to the very Koran, which does not contradict communism. If you go thus to a mullah and tell him, what communism means, and explain to him properly how to follow this idea, he will believe it saying: yes, it was written so long ago (Baberowski 2003, 277–278).

resonated, for example, in this famous Azerbaijani proverb: *Özünä 'happiness shared is a double happiness'* (('if you wish yourself one cow, wish your neighbour two'). This once again reflects the notions of 'good neighbourliness' and 'hamsoya' (Nurulla-Khodzhaeva 2021), as underpinning components of resilience-building in the country.

In this context, first, the historical background seems to be important. That is because people's expectations of what the desired living standards are remain largely shaped by Soviet practices, though welfare state was rolled back through a residualisation of social provisions during the post-Soviet time (Sayfutdinova 2015, 25). The *shadow of the (Soviet) past* looms still large: 'When the state does not fulfil those expectations, individuals and families do not simply give up on their expectations, but try to 'fill in the gaps' on their own to achieve the desired standards of living' (Sayfutdinova 2015, 25). This, in turn, puts forward the notion of self-help, emphasising people's reliance on their indigenous strengths/capacities over state aid or external help. One prominent example is a continued emphasis on higher education as a social value—an ideal, which feeds upon the Soviet practice of universally available education. It is no surprise that in all the focus groups countrywide, value to education was emphasised throughout the discussions.[25]

Considering the local, everyday level of being, Azerbaijanis are commonly people of 'hope', which make them generally—albeit in an irrational way—optimistic about the future even if the present looks rather bad than good. Thus, a (not rational, but rather faithful) belief prevails among people that the future will be positive, and everything is going to be alright in the end.[26] That is why people see everyday difficulties and negative experiences as temporary rather than permanent so that they often choose to look at them from the perspective of a metaphysical wisdom as cited in one of the focus groups (FGD 2019: 41-year-old female participant, college teacher in Shaki): *Yaman günün ömrü az olar* (bad times are short-lived).

At the local level, Azerbaijani society can be largely considered as 'traditionalist' (Sadigov 2014, 51). In particular, a narrow *familistic* orientation prevails within the communities. For instance, the overwhelming majority of people define their identity and, by extension, the meaning (purpose) of their life through the reference to family/kinship and their sense of wellbeing (CRRC Blog 2012).

A particularly child-centred familistic aspect of local communities strongly associates with the Azerbaijani society. Thus, an important element of people's *good life* concept in Azerbaijan refers to the (future) wellbeing of their children (and grandchildren).[27] Their future visions are therefore primarily not of individualistic but rather social nature, thus going beyond one's own personal happiness. For people, a most important part of their *good life* visions is having their own children, seeing them as being educated and married, and then having grandchildren (for example, FGD 2019, Shaki).

---

[25] The need to build up a high-quality education in the country, as well as seeing education as an essential part of the good life aspirations was observed unanimously in all the focus groups.

[26] This was particularly observed in the focus groups in Lankaran and Shaki (FGD 2019).

[27] In all the focus groups, the emphasis was laid that family is an important institution of human co-existence and is thus an indispensable part of good life aspirations.

Here again, marriage operates as a personally empowering mechanism that creates powerful social bonds within/across local communities. Because not only being born, but also being married into the 'right' family (e.g., a rich or high-status family) matters—as important part of one's future envisioning and/or planning: that is why marriage can be seen primarily as a very important tool for survival and thriving locally in Azerbaijan.[28]

Affective solidarity as a coping mechanism in local communities

In addition to an account of the above normative motivations and unified desires for social action, this article also argues that Azerbaijan society can be characterised largely by *affective solidarity*,[29] which in comparison with the so-called *rational solidarity*, is pre-rational and shorter, but is a stronger, more specific and immediate (aka local) form of social solidarity.

Affective solidarity can be caused by specific, local circumstances or events (things that make people, for instance, compassionate), and they are accompanied by high levels of arousal. Whereas people experience rational solidarity in stable, long-term and transcending social settings, they experience affective solidarity only when things are out of the ordinary or unusual in one's lifeworld (*Lebenswelt*). As such, they serve an adaptive role in helping them guide through their social behaviours in the face of various adversities in their immediate, local environment.[30]

It can therefore be argued that individuals' self-identification as members of their communities and the social bond between them are due to their social affects towards each other, common sentiments, values and beliefs as well. Here again, *affective experiences* of everyday life commonly help people stick and act together in a way that increases their chances of survival as part of their local communities. For example, a high level of social compassion in Azerbaijani society motivates people to help others in pain or need in their immediate social environment. Recent escalations in Karabakh testify to this: following three days of fighting on the Armenia-Azerbaijan border and the deaths of several Azerbaijani soldiers and officers, including an army general in mid-July 2020, a spontaneous and unorganised meeting of tens of thousands took place in the city centre to outpour rage about the unresolved Nagorno-Karabakh conflict and express solidarity and support for the country's military in a highly emotional way.[31] Equally, a 44-day war in the autumn of 2020 demonstrated an extreme level of unity in the society, when volunteers to fight

---

[28] As it was noted in one of the focus groups, a famous Azerbaijani joke shows, for example, how important marriage is in one's self-identification and rendering their purpose of life/future meaningful: *Where are you from? - I have not got married yet.*

[29] This concept parallels that developed by Durkheim (1984)—'mechanical solidarity': in his view, in traditional or small-scale communities solidarity is usually based on the social similarity of people (e.g. kinship ties). Likewise, *affective solidarity* constrains, but in more psychological fashion, people's natural egocentrism and causes them to have 'sympathy' and 'fellow feeling' towards other members of their community.

[30] For example, just as people run from wild animals due to fear caused as immediate emotion, they may try—in terms of affective solidarity—to help fellow members of their community being in need because immediately felt compassion elicits social affect.

[31] The chanted slogans were highly affective as well: 'Karabakh is ours', 'End the quarantine and start the war', or 'Karabakh or death' (Eurasianet 2020).

in the war formed long queues at the military commissions all over the country and the society mobilised resources for sending to the front in an unprecedented way. It also shows how affective solidarity can generate increased short-run popular support—the famous 'rally-round-the-flag' effect, because it is first and foremost people's patriotic sentiment in the face of dramatic events, boosting the collective feeling of belonging in terms of national identity.

At the local level, all focus group discussions (FGD 2019) attest to a more *relational* understanding of life and resilience, turning their attention to mutual interdependence and mutual responsibility within one's community: that is, Azerbaijani people tend to have a basis for 'empathy' built into their primary understanding of 'self'—with a strong social propensity for experiencing another's needs or feelings as one's own. Hence, they feel more comfortable/happier when related/connected to others and prefer relationship to separation.

Support infrastructures of the resilience in local communities

While people's self-identification including their good life aspirations as well as their solidarity constitute the ideational and emotional drives of communal resilience in Azerbaijan, a host of relevant, more tangible support infrastructures make it a living reality for people. What are the specific social infrastructures there to make it happen, and how do they operate? At the micro level, the Azerbaijani society can be characterised by being organised primarily around local communities of immediate kinship ties and neighbours; accordingly, Azerbaijanis' trust in family members, relatives, friends, and immediate neighbours is generally high, while they do not trust people they do not know (CRRC Blog 2012). For example, in almost all the focus groups, the participants emphasised that immediate family members (especially parents) were the first instance of support when in need (FGD 2019). Thus, the resilience of Azerbaijani communities is primarily based on such local forms of support infrastructure as networks of relatives or friends. Being already in place during the Soviet era, *bonding* social capital is historically quite prevalent in Azerbaijan. Individuals often survive and thrive (e.g., find jobs or gain access to resources) through family and kinship networks (see, for example, FGD 2019, Baku). Although Azerbaijani society has undergone a certain level of modernisation since the oil boom, as such it remains traditionally centred on personalistic, social (affective) solidarity institutions like family, kinship and neighbourhood (FGD 2019), which will be discussed below—as key examples for local support infrastructure of the individual.

Family

Like in many other traditional societies, in Azerbaijan, family is the most important institution that acts as the building block of the societal resilience. Family is the organising unit of social relations and trust, and sometimes even acts as the ultimate source of authority (CRRC Blog 2012). To rephrase Thatcher's famous comment, one can at the first sight even claim that there is almost no such thing as larger society in Azerbaijan, but only individuals and their families.

In societies like Azerbaijan, family is the context for self-perception, identity and values, and it also acts as a mechanism for protection and support (FGD

2019). The predominance of family and relatives as a mechanism of social relations is usually (but not exceptionally) characteristic for states with a weak welfare system, and its predominance in traditional societies, emanates from state weakness to a certain extent. In such states, certain primary functions of the state such as welfare, physical protection, adjudication etc. are carried out either by the extended family or by networks of kinship (Rotberg 2003). Predominance of family in the society as the main organising social unit may also leave a decreased room for individual freedoms and emancipation.[32] In such societies, individual decision-making and self-perception is often chained until one reaches a mature age, and on certain occasions, even further. And the dominance of the family over the individual is partly due to economic dependence and partly due to the role of customs and traditions.

As economic development unfolded in Azerbaijan through the windfall of oil and gas sales in the last decades, one could observe slight changes in individual emancipation that came about due to increased economic independence of individuals; this, however, has not changed the central role of family as a bonding institution, but rather materialised the leverage of families as local wealth and power centres over the individual (FGD 2019, Baku). Thus, economic progress, improvement in the state capacity and modernisation in Azerbaijan have hardly changed the traditional role family has played in the Azerbaijani society. This has been so partly because economic progress in the country has not been evenly distributed, and many individuals' lives have remained unaffected (Barrett 2020); as FGD (2019) confirms, this is particularly true in the rural areas of the country, which have had quite a different development from that of Baku. That said, even in Baku, and within the circles that have benefitted immensely from the economic boom, family has remained the dominant institution that restrains and determines individual identities, perceptions and decisions (FGD 2019, Baku).

Family has also played an important role in the shaping of social (including national) identity. As one focus group participant concisely put it, 'identity starts in the family' (FGD 2019: 47-year-old male participant, journalist in Baku). On the one hand, family itself and its preservation is the ultimate meaning of identity for most people, but on the other hand, the central role of family in the lives of the Azerbaijanis shapes national and social identity, leaving little room for individual identifications. The role of the family in defining one's social position is even more explicit, since in societies like Azerbaijan the social status is largely defined by which family you come from. Also, as one focus group participant emphasised, 'a first environment for the individual's self-perception and development, family plays a crucial role in defining his/her social behaviour' (FGD 2019: 40-year-old female participant, civil servant in Baku). Thus, the role of family in the resilience-building in the Azerbaijani society occupies a large space, and it does so by being a source of financial and moral support, protection as well as an institution that secures the reproduction of traditional values and customs in the society.[33] In times of global challenges and hardships (such as war,

---

[32] In some of the focus groups, both in Baku and outside, some of the participants confirmed the role of the parents in the making of essential decisions, sometimes limiting the individual's decision-making freedom (FGD 2019).

[33] For example, in most focus groups, the participants mentioned that in times of hardships in the last 30 years, the individuals had experienced resilience through the family institution (FGD 2019).

pandemic, etc.), family has been the first point of reference: as a most available social institution, it is one's family that helps them to self-organise and respond to such adversities. Therefore, there are strong family bonds, and individuals have a strong bond to parents even when they set up their own families. In this regard, the central role of the family in the Azerbaijani society is not only conditioned by relatively slow economic and social modernisation or relative state weakness, but more importantly, by traditions and norms that have survived the economic modernisation. So, family is a must in the Azerbaijani society, and it is largely taken for granted that every individual's ultimate goal is to have one. In that regard, as one FG participant put it metaphorically, 'families in Azerbaijan are like homes in blizzard, and anybody left outside will be swept through' (FGD 2019: 68-year-old male participant, pensioner/former school headmaster in Ganja). Solidarity within families is very high, and often intra-family solidarity is achieved at any price (FGD 2019); individuals may often seek to achieve what is good for their immediate families at the cost of the larger society, an example that one may encounter in Azerbaijan from time to time is when a driver partly blocks a public road by car when waiting for immediate family members (FGD 2019, Baku). Therefore, any solidarity that penetrates the local community or society comes through the immediate or the extended family, and is thus affective mostly, i.e., pre-rational and/or mechanical. Overall, family is a transmission channel, a conduit between the individual and society in Azerbaijan.

Kinship

Another important institution with even broader social repercussions is kinship that has had an impact upon the resilience-building of the local communities in Azerbaijan. When talking about kinship, the reference is made both to the immediate and distant relatives, that form an important support pillar for the reproduction and adaptation of the community. Kinship functions as a support infrastructure in two ways: as supplementary to the immediate family or as substitute when the immediate family members cannot provide support. It is especially in the latter form that kinship (relatives) contributes to the resilience-building of the society.

Relatives play an important role in supporting each other both financially and morally.[34] That means that even if they do not always act as a part of the material support infrastructure in Azerbaijan, the institution of relatives is still a very important one in non-material terms.[35] Therefore, it is often believed that the larger the kinship network is, the better chances are for individual survival and progress (FGD 2019).[36]

---

[34] For example, according to FGD 2019, most participants, especially outside Baku, mentioned that they felt morally obliged to support their poorer relatives financially.

[35] It was observed in the focus groups that most participants viewed kinship as an extension of the immediate family.

[36] As it was mentioned in some of the focus groups, kinship is seemingly perceived variably within the society. Some, especially the older generation attach a value to kinship that is closely related to survival, progress and material well-being. It was also mentioned that the younger generation attaches a different meaning to kinship in the light of the social changes undergoing through economic development.

Another dimension of kinship in Azerbaijan is the one related to communal (affective) solidarity. Like in the example of family, strong kinship institutions shape the development of social solidarity, by functioning as supplementary/ alternative to state institutions to a certain extent. One consequence of robust kinship relations is the formation of local solidarity (with a strong affective element) among relatives. Solid social bonds between relatives have helped them overcome many challenges in hard times. For example, many families in Baku have had the moral obligation of supporting their relatives who had become internally displaced people due to the war in the early 1990s (FGD 2019). On its turn, this affective solidarity of the micro-level can transform into more rational or larger societal solidarity, such as the increased tendency to support the socially vulnerable families during the COVID-19 pandemic of 2020–2021, when a lot of individuals volunteered in either financially contributing to or delivering aid to the relevant families. Equally, during the 44-day war in late 2020, there was an unprecedented level of mobilisation of resources and effort in Azerbaijan for supporting the soldiers in the front.

Neighbourhood

Neighbourhood is another important institution of the support infrastructure of resilience in Azerbaijan. It comes as third in line in addition to family and kinship, but can sometimes be stronger than the latter or replace it in the case of its absence (FGD 2019, Baku).[37] Neighbourhood has been an important dimension of social relations and sustenance in Azerbaijan for quite a long period of time. Neighbourhood, be it in the form of apartment blocks, or private houses or community housing had been an important institution of social life as well. Neighbourhoods have also played the role of social capital and can thus be easily utilised for collective action. For example, by organising themselves into social units, neighbourhoods played a key role in mass protests during the independence movement in the late Soviet Union, but also during the war with Armenia in its aftermath.

Moreover, neighbourhood has played a key role in the survival of the community at times of crisis such as wars, post-Soviet early transition related poverty, etc. For example, in a focus group in Baku, it was emphasised how individual households had relied upon each other in times of economic hardship in the immediate post-independence period (FGD 2019, Baku), and in more recent times of the health pandemic. Local practices such as lending money to people in need or putting regularly a certain amount of one's income into a 'neighbourhood pot' can be seen as a common example for mutual support mechanism among neighbours.[38] Thus, the concept of neighbourhood is very much an outcome of Azerbaijani culture of helping each other and is foundational for resilience-building in Azerbaijan, embedded in culture and traditions.

Unlike family and kinship institutions, neighbourhood's linkage to the broader society or societal solidarity is more direct and straightforward. As

---

[37] In general, the famous proverb 'Better a neighbour nearby than a relative far away' applies here strongly.

[38] Then, each month, a contributing family to the 'neighbourhood pot' takes the whole sum from it to cover their large expenses, which they could not have done otherwise.

such, unlike family and kinship, neighbourhood has not produced nepotism, keeping however all the other features of the support infrastructure.[39] For example, in good times or bad, neighbours have supported each other by helping financially, providing moral support as well protection. However, it should be noted that the neighbourhood concept already existed and was practised in the old Soviet era buildings and settlements of Baku and has not moved to the new skyscrapers or apartment complexes. Equally, the concept of 'məhlə', which is a form of self-identification and social unit for the younger male population, exists in the old neighbourhoods of Baku.[40]

In recent years, neighbourhood has been challenged by several social processes, especially in Baku. Neighbourhood as a concept has, to some extent, lost the previous significant role it had played in the resilience of the community during the Soviet and immediate post-Soviet periods. First and foremost, a number of neighbourhoods have ceased to exist because people have moved to newer and more modern urban complexes. Second, several tightly organised residential areas have been knocked down in Baku. Third, free market economy promoting more self-interested, individualistic values have changed the social relations between individuals (FGD 2019, Baku). Therefore, the neighbourhood concept in cities outside Baku and other regions have remained more intact because of the relative absence of the above-mentioned factors.

Azerbaijani peoplehood in the face of adversity

While identity (ideationally) and support infrastructures (socially) keep community in existence/together, peoplehood describes a community's ability to effectively mobilise their (often scarce) resources required for success in the face of adversity. Thus, here it is a useful way to refer to peoplehood as a *community in (collective) action* in times of trouble (Korosteleva and Petrova 2020). The concept shows how a community is being rendered into peoplehood—by their members be-coming together to meet challenges or crisis.

To this end, Azerbaijani peoplehood is a highly mobilised (organised) entity. The resilience of the Azerbaijani society derived historically from the organisational demands of various challenges, and people's resilient identity owes its origins and development to a large degree to the effects of earlier adversities. Historically, the modern Azerbaijani society with its highly resilient character was born during a time of unprecedented crises—particularly in the context of an enduring ethnic/national rivalry with Armenians starting in the early twentieth century.[41]

Various factors serve as a social-psychological referent in creating a sense of community in Azerbaijan. Historically, it has become strong throughout the

---

[39] It was mentioned in one of the focus groups in Baku that unlike kinship, neighbourhood entails less or no material element.

[40] *Məhlə* is mostly used by the younger male population to refer to a specific form of neighbourhood. As such, the concept refers to much more than mere neighbourhood, since it entails some form of local ownership and community. In particular, in the old neighbourhoods of Baku, there is the concept of honour of *məhlə*.

[41] In all the focus groups, the biggest external threats were said to emanate from Armenia, especially through the latter's occupation of Azerbaijan's territory (FGD 2019). It was also emphasised that this external threat has shaped the nation and made it stronger.

twentieth century and has gained a new momentum in the post-independence period. Factors that have shaped the peoplehood include identity, common values, role of the intelligentsia, especially of 1930s as well as Soviet policies of supporting the cultural autonomies of the union republics. In recent history, Azerbaijani people's *shared experience* of the two highly dramatic events that occurred concurrently has been especially significant in understanding the strong sense of peoplehood and its *resilient* character in the country: the collapse of the Soviet Union and the first Karabakh War—both taken place in the early 1990s. In a Nietzschean sense ('What does not kill you makes you stronger'), out of 'life's school' of struggle—these adversities did not 'kill' Azerbaijani sense of community, but rather made it stronger.

In the post-independence period, Azerbaijani peoplehood has thus acquired new impetus through mostly the difficult path the country has gone through. The independence movement, the war and especially the relatively larger number of human costs on the Azerbaijani side, atrocities such as the Khojaly massacre of 1992, political instabilities and occupation of a serious portion of the territory by the neighbouring Armenia as well as influx of a large number of internally displaced persons and refugees, added by socio-economic problems in everyday life such as systemic corruption and relative deprivation have all played a role in shaping the sense of peoplehood.

How does daily life make people stronger in the face of adversity? For example, in the context of the long-standing Nagorno-Karabakh conflict, the sense of respecting the fallen soldiers as martyrs and their close relatives as martyrs' families, the need to provide protection for them have become integrated into the social support infrastructures contributing to the resilience of the Azerbaijani society. Thus, the conflict has on the one hand played a decisive role in shaping the national identity, and on the other hand strengthened locally the sense and solidarity of community.

In terms of mobilisation of the community into a peoplehood, Azerbaijani people with their communal self-reliance and solidarity tend to move away from the Soviet-time mentality of state dependence and/or external aid, with an emphasis on the local resources and infrastructures as more attuned to meeting their needs. It is, for example, a common practice in the country that people help each other financially during—privately organised—large and expensive events (such as wedding or mourning ceremonies) in their community: that is, all participants in such ceremonies give the host cash payments, the amount of which can vary—besides reciprocity—depending on participants' financial position or their social distance to the host.

Another important aspect of 'togetherness' is the people's aspirations for a good life as a 'pull factor' of resilience-building, by striving to build happier and stronger relationships, including communities. In all the focus groups conducted in Baku and other locations of Azerbaijan, there was an emphasis on a future-oriented understanding of resilience (FGD 2019). The good life aspirations were partly material and partly ideational, related to individual happiness, self-determination, but also issues such as cessation of the war and reintegration of Nagorno-Karabakh, building a more affective civil society as well as raising responsiveness of government institutions to the society's concerns.

Finally, it can be argued that the Azerbaijani peoplehood are capable and resilient enough to meet global (and/or international) challenges effectively at the local level and to respond accordingly to the forces of globalisation. In terms of the effect of global governance, globalisation is seen both as an opportunity and challenge in Azerbaijan.[42] On the one hand, globalisation means to many people, access to information, better education, getting to knowing other cultures; but on the other—weakened national values and traditions (e.g., family values). In this respect, developing a strong national educational system that would bind together the younger generation around values and education may be conducive building resilient societies that both embrace tradition and look to the future.

Conclusion

'Going local' matters indeed in international affairs since 'the global' both begins and ends with 'the local'. Nothing could illustrate more vividly the distinctive global/local aspect of our increasingly complex world. The main effects of its global forces and drivers invariably fall on the local level. But conversely, local events and practices can also have profound global implications. They serve to animate and link all the other levels of today's (self-)governance. Here, then, the question arises of how people locally self-organise in practice and become more resilient?

The case of Azerbaijan shows the characteristic way in which 'the local' generates societal resilience in times of a complex change, and what role indigenous (self-)empowerment, especially communal solidarity and local cultural values play in resilience building. In doing so, this paper has referred to Azerbaijani people and society as the social-psychological entity of a special sense of both past- and future-oriented identification with the community: in this view, community members are the 'people' of one's ancestors, therefore they are one's people, and they will be the people of one's children and their children, and thus constitute a so-called *covenant of fate and destiny*. As such, it also entails responsibility for the future of Azerbaijani people and their community/-ies. Thus, it can be concluded that people's self-identification/resilience is genuinely *relational*, meaning that they depend on each other for their (search for a) meaning of life.

In particular, this paper has addressed the development of a sense of community in Azerbaijan based on people's identity and aspirations for 'the good life', which in turn are being supported through formal and informal communal infrastructure such as family or neighbourhood. In this way, Azerbaijani peoplehood primarily implies people's solidary bonds, or a sense that the community members share something more than their material well-being and prosperity. And here, the affective dimension of solidarity with a strong emotive element stands out among local practices in building people's independent indigenous capacity.

How resilient is Azerbaijani society? Resilience is all about the aspirations of a community, especially towards its long-term development. It can neither

---

[42] In most focus groups, globalisation was seen as an opportunity, however it was also emphasised that there is need to preserve the local dimension, while relating to the global level.

be defined nor implemented externally, because communities may develop varying understandings of it, and should oversee their own destiny, in terms of sustainable bottom-up, local empowerment (Korosteleva 2020; Chandler 2021). So, the good life aspirations of communities in Azerbaijan are also all about the choice of a life that one has reason to value, and resilience in Azerbaijan is largely about people's understanding that they are their own agents for building a better life. This paper has argued that the Azerbaijani society is rendered highly resilient by a combination of temporal and social dimensions: its (temporally) future-oriented and (socially) affective ways of workings significantly enhance the resilience of its people.

Finally, what the case of Azerbaijani local resilience means for global governance? While not directly offering ways to reshape international cooperation, it nonetheless shows what inner strengths and capacities can be engaged locally to address the challenges of complexity, which could serve as a starting point for the revision of regional and international engagement. Although not seeking to reconnect the local and the global, this article thus takes the first and necessary step to understanding the local and its (particular) importance in addressing the challenges of an increasingly complex world, which could further be explored in future research.

Disclosure statement

No potential conflict of interest was reported by the authors.

References

Altstadt, Audrey. 1992. *The Azerbaijani Turks: Power and Identity Under Russian Rule.* Stanford, CA: Hoover Institution Press.

Altstadt, Audrey. 2017. *Frustrated Democracy in Post-Soviet Azerbaijan.* New York, NY: Columbia University Press.

Archer, Margaret S. 2013. "Solidarity and governance" In *Governance in a changing world: Meeting the challenges of liberty, legitimacy, solidarity, and subsidiarity*, Extra Series 14, 1–2. Vatican City: Pontifical Academy of Social Sciences.

Auch, Eva-Maria. 2004. *Muslim—Untertan—Bürger: Identitätswandel in gesellschaftlichen Transformationsprozessen der muslimischen Ostprovinzen Südkaukasiens (Ende 18. bis Anfang des 20. Jahrhunderts)*. Wiesbaden: Reichert Verlag.

Babajew, Aser. 2010. "Zur Problematik von Nation und Nationalismus in Aserbaidschan." In *Jahrbuch Aserbaidschanforschung. Beiträge aus Politik, Wirtschaft, Geschichte und Literatur*, edited by Mardan Aghayev and Ruslana Suleymanova, 79–111. Berlin: Verlag Dr. Köster.

Baberowski, Jörg. 2003. *Der Feind ist überall: Stalinismus im Kaukasus*. München-Stuttgart: DVA.

Barrett, Tristam. 2020. "Your Debts Are Our Problem: Politicisation of Debt in Azerbaijan." *Focaal—Journal of Global and Historical Anthropology* 87: 1–15.

Berenskoetter, Felix. 2011. "Reclaiming the Vision Thing: Constructivists as Students of the Future." *International Studies Quarterly* 55 (3): 647–668.

Bourbeau, Phillippe. 2018. *On resilience: Genealogy, Logics and World Politics*. Cambridge: Cambridge University Press.

Bousquet, Antoine, and Robert Geyer, eds. 2011. "Complexity and the International Arena." *Cambridge Review of International* Affairs 24 (1): 1–3.

Caucasus, Barometer. 2013. *Azerbaijan*. https://caucasusbarometer.org/en/.

Chandler, David. 2021. "Decolonising Resilience: Reading Glissant's Poetics of Relation in Central Eurasia." *Cambridge Review of International Affairs*, 35 (2): 158–175.

Cornell, Svante E. 2002. *Autonomy and Conflict: Ethnoterritoriality and Separatism in the South Caucasus—Cases in Georgia*. Uppsala University Department of Peace and Conflict Research. Uppsala: Uppsala Universitet.

Cornell, Svante E. 2011. *Azerbaijan Since Independence*. London: Routledge.

Croissant, Michael P. 1998. *The Armenia-Azerbaijan Conflict: Causes and Implications*. West Port: Praeger.

CRRC Blog. 2012. *Trust and Agency in Azerbaijan: Personal Relationships versus Civic Institutions*, 13 Nov. 2012. http://crrc-caucasus.blogspot.com/2012/11/trust-and-agency-in-azerbaijan-personal.html.

Cureton, Adam. 2012. "Solidarity and Social Moral Rules." *Ethical Theory and Moral Practice* 15: 691–706.

Cusumano, Eugenio, and Stefan Hofmaier, eds. 2020. *Projecting Resilience Across the Mediterranean*. Cham, CH: Palgrave Macmillan.

de Beer, Paul. 2009. *Sticking Together or Falling Apart? Solidarity in an Era of Individualization and Globalization*. Amsterdam: Amsterdam University Press.

de Waal, Thomas. 2003. *Black Garden: Armenia and Azerbaijan Through Peace and War*. New York, NY: NYU Press.

Deutsch, Karl. 1972. *Nationenbildung—Nationalstaat*. Integration. Düsseldorf: Bertelsmann.

Donne, John. n.d. *John Donne's Devotions*. Grand Rapids, MI: Christian Classics Ethereal Library.

Durkheim, Emile. 1984. *The Division of Labour in Society*. London: Macmillan Press.

Ergun, Ayça. 2021. "Citizenship, National Identity, and Nation-Building in Azerbaijan: Between the Legacy of the Past and the Spirit of Independence." *Nationalities Papers*, 40–18.

Eurasianet. 2020. *Pro-war Azerbaijani Protesters Break into Parliament*, 15 July 2020. https://eurasianet.org/pro-war-azerbaijani-protesters-break-into-parliament

European Commission. 2020. *2020 Strategic Foresight Report*. https://ec.europa.eu/info/sites/default/files/strategic_foresight_report_2020_1_0.pdf

Focus Group Data (FGD). 2019. *Data Derived from over 20 Focus Group Discussions*. Conducted in the four major cities of the country (Baku, Ganja, Shaki, and Lankaran) in 2019.

Fischer Onar, Nora, and Kalypso Nicolaïdis. 2013. "The Decentering Agenda: Europe as a Post-Colonial Power." *Cooperation and Conflict* 48 (2): 283–303.

Goltz, Thomas. 2015. *Azerbaijan Diary: A Rogue Reporter's Adventures in an Oil-Rich, War-Torn, Post-Soviet Republic*. London: Routledge.

Hirose, Yoko. 2016. "The Complexity of Nationalism in Azerbaijan." *International Journal of Social Science* Studies 4(5): 136–49.

Hooker, Juliet. 2009. *Race and the Politics of Solidarity*. Oxford: Oxford University Press.

Human Rights Watch/Helsinki. 1994. *Azerbaijan: Seven Years of Conflict in Nagorno-Karabakh, New York.* https://www.hrw.org/reports/pdfs/a/azerbjn/azerbaij94d.pdf

Hunt, Scott A., and Robert D. Benford. 2004. "Collective Identity, Solidarity, and Commitment." In *The Blackwell Companion to Social Movements*, edited by David A. Snow, Sarah A. Soule, and Hanspeter Kriesi, 433–457. New York, NY: Blackwell Publishing.

International Crisis Group (ICG). 2004. *Azerbaijan: Turning Over a New Leaf?* 13 May 2004, Europe Report No.156. Baku/Brussels. https://www.crisisgroup.org/europe-central-asia/caucasus/azerbaijan/azerbaijan-turning-over-new-leaf.

Ismayilov, Murad. 2018. *The Dialectics of Post-Soviet Modernity and the Changing Contours of Islamic Discourse in Azerbaijan: Toward a Resacralization of Public Space.* Lanham: Lexington Books.

Ismayilov, Murad. 2019. "Islamic Radicalism that Never Was: Islamic Discourse as an Extension of the Elite's Quest for Legitimation. Azerbaijan in Focus." *Journal of Eurasian Studies* 10 (2): 183–196.

Kappeler, Andreas. 2001. *Russland als Vielvölkerreich: Entstehung, Geschichte, Zerfall.* München: C.H. Beck.

Kavalski, Emilian. 2007. "The Fifth Debate and the Emergence of Complex International Relations Theory: Notes on the Application of Complexity Theory to the Study of International Life." *Cambridge Review of International Affairs* 20 (3): 435–454.

Kavalski, Emilian, ed. 2016. *World Politics at the Edge of Chaos: Reflections on Complexity and Global Life.* Albany: State University of New York Press.

Korosteleva, Elena. 2020. "Reclaiming Resilience Back: A Local Turn in EU External Governance." *Contemporary Security Policy* 41 (2): 241–262.

Korosteleva, Elena, and Trine Flockhart. 2020. "Resilience in EU and International Institutions: Redefining Local Ownership in a New Global Governance Agenda." *Contemporary Security Policy* 41 (2): 153–175.

Korosteleva, Elena, and Irina Petrova. 2020. "From 'the Global' to 'the Local': The Future of 'Cooperative Orders' in Central Eurasia in Times of Complexity." *International Politics* 58: 1–37.

Korosteleva, Elena, and Irina Petrova. 2022. "What Makes Communities Resilient in Times of Complexity and Change?" *Cambridge Review of International Affairs* (forthcoming, as part of the Special Issue "The Making of Resilient Communities in Central Eurasia in Times of Complexity and Change." 35(2): 2022)

Melucci, Alberto. 1988. "Getting Involved: Identity and Mobilization in Social Movements." *International Social Movements Research* 1, 329–348.

Melucci, Alberto. 1996. *Challenging Codes: Collective Action in the Information Age.* New York, NY: Cambridge University Press.

Moreno, Alberto P. 2005. "The Creation of the Azerbaijani Identity and Its Influence on Foreign Policy." *UNISCI Discussion Papers*, May 2005. https://www.ucm.es/data/cont/media/www/pag-72533/Alberto8.pdf

Motika, Raoul. 2009. "Aserbaidschan-Nationalismus und aseritürkischer Nationalismus." In *Nationalismus im spät- und postkommunistischen Europa: Band 2—Nationalismus in den Nationalstaaten*, edited by Egbert Jahn, 299–329. Baden-Baden: Nomos.

Najafizadeh, Mehrangiz. 2012. "Gender and Ideology: Social Change and Islam in post-Soviet Azerbaijan." *Journal of Third World Studies* 29 (1): 81–101.

Nietzsche, Friedrich. 2013. *Götzen-Dämmerung: oder Wie man mit dem Hammer philosophiert.* Leipzig: Naumann Verlag.

Nordin, Astrid H. M., Graham M. Smith, Raoul Bunskoek, Chiung-chiu Huang, Yih-jye (Jay) Hwang, Patrick Thaddeus Jackson, Emilian Kavalski, L H M Ling, Leigh Martindale, Mari Nakamura, Daniel Nexon, Laura Premack, Yaqing Qin, Chih-yu Shih, David Tyfield, Emma Williams, and Marysia Zalewski. 2019. "Towards Global Relational Theorizing: A Dialogue Between Sinophone and Anglophone scholarship on Relationalism." *Cambridge Review of International Affairs* 32 (5): 570–81. DOI:.

RESILIENT COMMUNITIES OF CENTRAL EURASIA 151

Nurulla-Khodzhaeva, Nargis. 2021. ""Imitated" and Intuitional Resilience of Sufi-hamsoya." *Cambridge Review of International Affairs.* Unpublished manuscript (presented at the GCRF COMPASS WiPS)).

Romano, Santi. 1947. *Principii di diritto costituzionale generale.* Milano: Dott. A. Giuffrè.

Rorty, Richard. 1989. *Contingency, Irony, and Solidarity.* Cambridge: Cambridge University Press.

Rotberg, Robert I. ed. 2003. *State failure and State Weakness in a Time of Terror.* Washington DC: Brookings Institution Press.

Rouet, Gilles, and Gabriela C. Pascariu, eds. 2019. *Resilience and the EU's Eastern Neighbourhood Countries: From Theoretical Concepts to a Normative Agenda.* Cham, CH: Palgrave Macmillan.

Sadigov, Turkhan. 2014. "Corruption and Social Responsibility: Bribe Offers among Small Entrepreneurs in Azerbaijan." *East European Politics* 30 (1): 34–53. DOI:.

Safiyev, Rail. 2018. *Hinter der glitzernden Fassade: Über die Macht der Informalität in der Kaukasusrepublik Aserbaidschan (Edition Politik, Bd. 56).* Bielefeld: transcript Verlag.

Sayfutdinova, Leyla. 2015. "Negotiating Welfare with the Informalizing State: Formal and Informal Practices among Engineers in post-Soviet Azerbaijan." *Journal of Eurasian Studies* 6 (1): 24–33.

Scholz, Sally J. 2008. *Political Solidarity.* University Park: The Pennsylvania State University Press.

Shaffer, Brenda. 2002. *Borders and Brethren: Iran and the Challenge of Azerbaijani Identity.* Cambridge, MA: The MIT Press.

Tocci, Nathalie. 2020. "Resilience and the Role of the European Union in the World." *Contemporary Security Policy* 41 (2): 176–194.

Community resilience and social capital in post-Soviet mono-industrial areas affected by the uranium legacy and radiation: evidence from Kyrgyzstan

Chiara Pierobon,🆔 and Zarina Adambussinova 🆔

Abstract   *Throughout the entire Soviet era, the Central Asian region served as the main supplier of various mineral materials (including uranium and rare-earth metals) for the ambitious, Party-led political, military, economic, and social projects. The extensive mining and processing of radioactive ores resulted in a vast number of high-risk tailing dumps and other types of legacy sites across the region, thereby posing serious risks to public health and the environment. Our study focuses on a case in contemporary Kyrgyzstan to assess community resilience in formerly mono-industrial urban areas affected by uranium legacy and radiation. This chapter is based on an extensive review of the literature related to the concepts of community resilience and its relationship to social capital, and field research conducted in three post-Soviet monotowns: Ak-Tuz, Kadji-Sai and Orlovka. Using these data, we present empirical findings that will help explain what resilience means for the local communities and how it can be assessed within the context of the post-Soviet transition. Our study reveals some important findings about what makes some communities stronger than others, and what could enable more effective, sustainable interventions to support them further taking into account their local settings and social assets.*

Introduction

During the Soviet era, Central Asia served as the main supplier of uranium and rare-earth metals to the USSR for more than 50 years when uranium ore was being extracted and imported from different countries in the region for processing (EBRD, n/a). Consequently, a significant amount of mining waste, including radioactive waste, was generated and is still present at both former and currently operational mining sites throughout the region. This uranium legacy still presents a serious threat to residents' livelihoods, human health, and the environment as extant tailings continue to pollute groundwater, the atmospheric air (in dust, aerosols, vapours), and the biosphere (UNDP, 2019: 2). Most of these sites are located along the tributaries of the Syr Darya River, which runs through the Fergana Valley and the Kyrgyz Republic, Tajikistan, and Uzbekistan—an area that is both densely populated and agriculturally active (Reiserer 2021). In addition, after the break-up of the Soviet Union, these mining sites were largely abandoned. Because the Soviet-era documents that might contain relevant information about these sites are inaccessible, radioactive materials that are still present within non-operational mining facilities tend to be poorly secured, and it is still unclear whether other dangerous radiological materials exist at former mining sites (Humphrey and Sevcik 2009). This represents a particularly thorny issue for the Central Asian region, which has witnessed political instability and religious radicalisation with terrorist ties (Pierobon 2021a).

When Kyrgyzstan was still part of the USSR, the country was involved in various Socialist engineering projects. The territory of Soviet Kyrgyzstan in particular provided 'hundreds of mineral deposits of different types' (Bogdetsky et al. 2005, 6) through a web of 20 towns and villages that constituted the mining area (ibid.). Today, Kyrgyzstan is characterised as a 'fragile resource-dependent' country (Ocakli et al. 2020, 194) for which the mining industry remains the most vital resource within its economic realm. Recently, the gold-mining industry has become the mainstream of the country's local economy (ibid.). Kyrgyzstan is currently home to 92 hazardous waste dumps, which represents 475 million tons of waste that contains toxic substances (UNDP 2019). Tailings of radioactive waste account for 50.6 million cubic meters of this waste, and most of the uranium tailings, as well as the sites where heavy metals are extracted, are located in Central Asia (Ak-Tuz, Kadji-Sai, Kyzyl-Jar, Mailuu-Suu, Min-Kush, Orlovka, and Shekaftar) and in Toomoyun (OSCE n/a).

Since gaining its independence in 1990 after the collapse of the Soviet Union, Kyrgyzstan has undergone a crucial political shift toward reform and a transition to democracy, as evidenced by three pro-democracy uprisings since 2000. Today, 60% of the country's population (about 6.6 million in total) is located in remote mountainous regions. Unfortunately, local communities are still hardly aware of the risks associated with radiological waste and do not effectively engage in uranium risk management (UNDP 2019, 3). In the past, several programmes were implemented with the support of the international donor community aimed at addressing Kyrgyzstan's uranium legacy, and in response to United Nations resolutions A/RES/68/218 and A/RES/73/238 (in 2013 and 2018, respectively), which acknowledged the role of the international community in preventing the radiation threat to the region. The largest and most relevant project in this field is represented by the Environmental Remediation Account for Central Asia (ERA). Established in 2015, the ERA initiative is led by the European Union (EU) and is managed by the European Bank for Reconstruction and Development (EBRD) (Reiserer 2021). Among the former sites of uranium production, the EBRD has identified seven as having the highest priority in addressing the current situation: Mailuu-Suu, Min-Kush, and Shekaftar in Kyrgyzstan; Degmay and Istikol in Tajikistan; and Charkesar and Yangiabad in Uzbekistan.

In parallel with infrastructural initiatives aimed at recultivation and rehabilitation, community-level projects have been implemented by organisations such as the Organization for Security and Co-operation in Europe (OSCE) and the United Nations Development Program (UNDP), which are focused on 'reducing the risk of the negative impact of the uranium waste on people, livelihoods, and environment through raising awareness and supporting people-centred, gender-sensitive, risk-informed solutions in legacy sites at the level of local communities' (UNDP 2019, 1).[1] A look at their programmatic documents reveals that a major objective of these initiatives has been to 'build resilience to shocks and crises through enhanced prevention and risk-informed development' (ibid.).

Over the past two decades researchers and practitioners have increasingly emphasised the importance of fostering community resilience in response to both natural and man-made stressors, shocks and disasters. Although interest in such efforts has led to a proliferation of tools to assess community resilience in the face of disasters, 'learning from existing levels of resilience among communities such as residents in informal settlements has aroused limited attention' (Vertigans and Gibson 2019, 624). Indeed, previous efforts concentrated mainly on the 'multitude of environmental, economic and social problems that many communities share and [the] internationally derived solutions' (ibid.). Following Vertigans and Gibson's example (2019), we have addressed this gap by exploring the concept of resilience that characterises mining sites affected by the uranium legacy and environmental contamination in Kyrgyzstan. To do so, we adopted a social capital lens and focused on the experiences and perceptions of communities on the ground. As a matter of fact, research on resilience has yet to fully embrace social capital as a critical component despite the fact that evidence about its usefulness is already available. By assessing the community resilience and the role of social capital in former monotowns, we can shift our attention to the local level, putting communities at the centre of the analysis as a way to understand 'why some communities stay more resilient than others, even if they may have fewer resources and be less prosperous comparatively in material wealth' (Korosteleva and Petrova 2022, 7; and in this volume). Our findings shed light on how to design more effective and sustainable interventions that are targeted to local settings and that make use of the available social assets.

At the outset, this chapter will familiarise the reader with the term 'resilience' and explain its relationship to social capital. We then briefly present case studies of three post-Soviet monotowns: Kadji-Sai and Orlovka and Ak-Tuz considered as one Soviet industrial complex. The section on methodology provides detailed information regarding the model we used for assessing resilience as well as our data collection process. Next, we present the empirical findings, which are based on qualitative, semi-structured interviews and focus-group discussions (FGDs) that were conducted in the target locations and in Bishkek, the capital of Kyrgyzstan. Finally, we reflect on the relevance and limitations of the concept of resilience and social capital theory in post-Socialist transition societies and identify possible directions for future research in this field.

Community resilience and social capital
Although the concept of community resilience remains abstract and has been used as a buzzword by social researchers, practitioners, and policy-makers, resilient attitudes and behaviours have been increasingly framed within the Global North as a potential approach to meeting community developmental challenges (see Steiner and Markantoni 2013, and Vertigans and Gibson 2019). Since the late 1990s, international donors have also aligned with this trend by more frequently insisting on the centrality of 'the local' (Ejdus and Juncos 2018). For example, the EU Global Strategy of 2016 put forward the argument that 'positive change can only be home-grown', and it highlighted the importance of stimulating resilience through 'locally owned rights-based approaches' and by blending top-down and

bottom-up efforts to foster local agency (European Union 2016). Nonetheless, in the past decades, several studies have pointed to the gap between the rhetoric and the practice of international donors in this field. For instance, for Petrova and Delcour (2020), the EU's broad conceptualisation of resilience and local ownership has resulted in a narrow operationalisation in terms of policy instruments and in the use of previously established policy templates and mechanisms on the ground (Petrova and Delcour 2020, 353). Similarly, Korosteleva has criticised the EU resilience-thinking strategy for being too outside-in 'by way of offering external solutions to internal problems for communities, turning them into dependable subjectivities and consumers of the western modes of "good governance"' (Korosteleva 2020, 253). It is in this context that a new understanding of resilience was introduced as 'an analytic of governance, that focuses on developing the internal strength and capacities of a system' (ibid., 244). Accordingly, communities are seen as having 'capacities and coping strategies that are more attuned to resolving the problems on the ground, also with external support if necessary' (Korosteleva and Petrova 2022, 3, and in this volume).

Community resilience can thus be conceptualised as 'the collective ability of a neighbourhood or geographically defined area to deal with stressors and efficiently resume the rhythms of daily life through cooperation following shocks' (Aldrich 2012). The literature also refers to resilience as the capacity to 'respond to and influence change, to sustain and renew the community and to develop new trajectories for the communities' future' (Magis 2010, 402). Remarkably, a diverse set of capacities is necessary to meet both the reactive and the proactive challenges posed by economic, political, environmental, and social shocks (Bernier and Meinzen-Dick 2014, 2). In this framework, Keck and Sakdapolrak (2013) have identified three types of capacities for resilience: coping capacities, adaptive capacities, and transformative capacities. *Coping capacities* refer to resources that are directly available to people to overcome immediate threats (Keck and Sakdapolrak 2013, 10). *Adaptive capacities* correspond to people's ability to anticipate future risks and adjust their livelihoods accordingly by employing what they have learned from past experiences (and those of others) (ibid.). *Transformative capacities* refer to:

> people's ability to access assets and assistance from the wider socio-political arena (i.e., from governmental organizations and so-called civil society), to participate in decision-making processes, and to craft institutions that both improve their individual welfare and foster societal robustness toward future crises.
>
> (Keck and Sakdapolrak 2013, 11).

Coping, adaptive, and transformative capacities for resilience are grounded in the intra- and inter-relations that constitute the 'relational communal infrastructure' (see Korosteleva and Petrova 2022, in this volume) through which resources become available and/or can be accessed. Scholars have recently pointed out the centrality of this network of relations – that is, social capital – for building both individual and community resilience (Kwok et al. 2019). *Social capital* can be described as the 'civic engagement networks, or reciprocal and trusting human relations based on shared norms' that function as a 'catalyst[s] for local action' (Jacobs and Hofman 2019, 403). It comprises both local synergies (also known as *bonding social capital*) as well as 'local actors' abilities to reach out to external resources' (also known as *bridging or linking social capital*) (Jacobs and Hofman 2019, 403). For scholars such as Bernier and Meinzen-Dick (2014), it is possible to draw parallels between the three coping, adaptive, and transformative capacities for resilience and the bonding, bridging, and linking types of social capital.

Bonding social capital refers to strong ties within homogeneous groups based on location, ethnicity, religion, shared values, and the like that sustain intragroup cohesion and integration. By providing access to resources that are directly available within groups, this form of social capital increases the coping capacities of target communities. *Bridging social capital* refers to ties between actors and groups across social stratifications and identities that tend to be 'dissimilar' with respect to features such as socio-economic status, ethnicity, and so on (Hawkins and Maurer 2010, 1779–80). Because of this dis-homogeneity, the network created is characterised by weaker linkages (Granovetter 1983) but – most importantly for developing adaptive resilience capacities—provides access to new skills and resources otherwise not available (Onyx and Bullen 2000). As a matter of fact, through bridging social capital, individuals and communities may be more effective in obtaining support to solve problems or cooperate on a wider spatial scale (Jacobs and Hofman 2019, 403). *Linking social capital*, a peculiar form of bridging social capital, refers to ties between individuals who occupy different positions of authority across vertical power lines. This form enables the development of transformative resilience capacities in that it offers access to power structures and institutions through which these individuals can influence policy (Woolcock and Narayan 2000). Network brokers play a pivotal role in the creation of adaptive and transformative capacities for resilience (see also Morgan-Trimmer 2013) because they are capable of activating bridging and linking social capital and of overcoming the downside of (bonding) social capital.

Kadji-Sai, Orlovka, and Ak-Tuz as post-Soviet monotowns

A *monogorod* (or a mono-industrial town/monotown) is a settlement formed around one enterprise or a group of enterprises that are connected by a single production chain. The term is frequently articulated in official and public discourses in different post-Soviet countries (Kazakhstan, Russia, and rarely in Kyrgyzstan) and refers to the post-Soviet mono-industrial urban areas characterised by a high rate of unemployment, de-urbanisation processes, environmental problems, and health issues.[2] Although one can date the establishment of the first monotowns in Central Asia back to the 1930s, to a large extent industrial towns were widely in evidence by the 1950s and 1960s and were supported by Moscow-led provisioning (*moskovskoe obespechenie*). In the past, residents of Soviet monotowns would commonly benefit from a range of privileges organised around a town-forming enterprise. The industry acted as the key provider of such benefits, such as a higher-than-average salary, land for building one's own house or apartment, newly built residential houses with running water and central heating, and an advanced city infrastructure and services (Nasritdinov, 2015; Junussova and Beimisheva 2020). The collapse of the USSR led to dramatic changes in the monotowns and the lives of their residents, turning most of these settlements into economically marginalised areas in the 1990s and isolating the communities of citizens who remained there. The current residents continue to struggle with numerous pressing issues, including the complete or partial shutdown of town-forming enterprises, high unemployment resulting in impoverishment and outmigration, de-urbanisation (decaying Soviet-era industrial buildings and housing), widespread insecurity, and inadequate basic communal services.

In contemporary Kyrgyzstan, a number of post-mining towns and villages serve the metallurgy, chemical, and coal industries. Former mining towns receive little scholarly attention, especially with regard to local communities and their concerns. In most academic accounts, the dominant discussion relative to the mining industry is limited to analysing different aspects of the protests and conflicts associated with (gold) mining (e.g., Bogdetsky and Novikov 2012; Doolot and Heathershow 2015; Moldalieva and Heathershaw 2020; Ocakli et al. 2020). In the current study, we explain the idea of a monotown and attempt to assess community resilience in such urban areas. Since the mid-1990s Kyrgyzstan's government has articulated the importance of assistance from the international community when it comes to enhancing radioactive security and ensuring long-term sustainability in such areas (e.g., from the 'Environmental Action Plan [NEAP]' [1995−97], the 'Programme of Cooperation on Combating the Smuggling of Nuclear and Radioactive Materials' [2007], and a regional conference organised by UNDP on 'Uranium Tailings: Local Problems, Regional Consequences and Global Solutions' [2009]).

This paper focuses on former monotowns that were previously operated for the mining of uranium and heavy metals. Today, most of these towns receive regular financial aid through various top-down international projects, some of which aim to build community resilience. However, these solutions seem to ignore the specific design of such urban areas (their peripheral position geographically, labour sourced from the whole Soviet Union, direct subsidies from Moscow) and the dependence of the local economy and social life on the earlier town-forming industries. These conditions create significant barriers not only to diversifying local livelihoods in the post-mining period but also to strengthening community resilience.

Kadji-Sai village in the Ton district of the Issyk-Kul region, with its population of 4,456 (2018), has one uranium tailing with a volume of 0.4 million cubic meters. The uranium waste dump is located about 180 metres above the Issyk-Kul Lake and 2.5 km from the populated area of Kadji-Sai known as 'Promka', the former industrial zone 3 km from the downtown area. As a typical labour settlement, Kadji-Sai was founded in 1947 on the southern coast of the lake at the height of 1,979 meters above sea level. Uranium exploration and extraction in Soviet Kyrgyzstan began in 1943 as part of the Soviet military policy and its developing nuclear weapons program. Uranium oxide in Kadji-Sai was mined from ashes of brown coal that contained uranium from a mine of the Sogutin deposit that was in operation from 1949 to 1966/68. For further processing, production was then undertaken by the Kara-Balta Mining Enterprise.

Uranium mining activities in Kadji-Sai ceased at the end of the 1960s owing to low profitability. Since that time, coal extraction became the key industrial activity in Kadji-Sai and supported the livelihoods of the inhabitants until the 1990s. Given its status as a priority site affected by the uranium mining operations in Kyrgyzstan, the village is constantly included in different international projects, which are financed mostly by UNDP, EBRD, and Rosatom and are aimed at enhancing 'awareness-raising and public outreach' (UNDP 2019, 4). Examples of such projects include Phase I of 'Socio-economic development of communities around radioactive sites in Kyrgyzstan' (2014) and the Commonwealth of Independent States [CIS] 'Remediation of territories of states affected by uranium mining operations' (2013–2023).

Present-day Ak-Tuz is a former single-enterprise village of 800 people, including labour migrants.[3] It is currently surrounded by 2.3 million cubic meters of radioactive waste left in four tailings that were developed between 1942 and 1978 in close proximity to the populated area. Orlovka has a population of 6,167 people (2021), and in 2012 it was officially deemed a town of *rayon* significance in the Kemin district of the Chui region. As a result of Soviet mining activities at the Bordu deposit, a large tailing can be found about 3.8 km from the town. This tailing contains 3.2 to 3.7 million cubic meters of radioactive waste and occupies an area of 130,000 cubic meters. The large-scale mining and processing activities that took place in both these locations resulted in a huge amount of accumulated industrial and toxic wastes that contain radioactive thorium and salts of heavy

Table 1. Examples of the resilience-related questions for investigating the environmental dimension

---

1. Do you have any problems related to:
   - The quality of air?
   - The quality of water, such as heavy metals in water?
   - Changes in the earth, such as landslides and land degradation?
   - The environment in general, such as radiation?
   - Or changes affecting the weather?
2. Are you usually able to overcome these problems by using the resources and knowledge that you have available?
3. Do your family and neighbourhood support you in dealing with these problems? If yes, how?
4. Do local and national authorities support you in dealing with these problems? If yes, how?
5. Do you receive support from the international community/international organisations in this field? If yes, how?
6. Do you receive support from local NGOs or other social groups? If yes, how?
7. Are you a member of any groups/NGOs that deal specifically with these environmental issues?
   - If yes, which group and what do you do? Which are your target groups, etc.?
8. Have there been any changes in the ways in which you deal with these problems compared to the past?
   - If yes, what have you learned and from whom? Who provided you with new knowledge and information?
9. Have you been involved in any official discussions or initiatives concerning these problems?
10. What kind of assistance is still needed in this field?

Source: Authors' own compilation.

metals (cadmium, molybdenum, lead, zinc, beryllium, hafnium, and zirconium oxides). In Ak-Tuz, there is currently no mining industry or any other industrial activity. In Orlovka, since 2015, a Chinese gold-mining enterprise known as Altynken LLC[4] has been officially operating at the working mine called Taldy-Bulak Levoberezhny 12 km from a town located along the Taldy-Bulak River.

Methodology

This exploratory study relied on an extensive literature review as well as qualitative field research. Measuring resilience is challenging since no available tool has been universally agreed upon. In our study, we opted for a qualitative approach and elaborated our own analytical framework for understanding and analysing community resilience by focusing on existing environmental, economic, and social challenges as described by *real people* in former mono-industrial towns in the peripheral areas of Kyrgyzstan. More specifically, community resilience was analysed in terms of the capacities of local communities to deal with these challenges. These capacities were investigated by looking at the type of resources that individuals can access through their social capital—that is, social ties, networks, and linkages to other individuals, groups, and institutions.

To evaluate resilience in qualitative terms, we constructed a questionnaire for both the Focus Groups (FGDs) and the semi-structured interviews that involved community members at the research sites. The questionnaire was divided into three large sections designed to cover the main environmental, economic, and social challenges that are faced by contemporary monotowns. First formulated in English, the questionnaire was then translated into Kyrgyz and Russian. Each section of the questionnaire comprised 10 questions that were constructed around the three capacities for resilience (coping, adaptive, and transformative) and the three related dimensions of social capital (bonding, bridging, and linking). (Table 1 gives examples of questions in the section on environmental challenges.)

Fieldwork was conducted in three former mono-industrial areas in the northern part of Kyrgyzstan: Kadji-Sai (November 2021), Orlovka (December 2021), and Ak-Tuz (January 2022). Overall, 12 FGDs were held both in Kyrgyz and in Russian with six groups of residents in Kadji-Sai and Orlovka, including youths, women and men of different ages and professional backgrounds, and local activists. The number of participants ranged between 6 and 12 per FGD; overall, 87 people were involved in the FGDs, and each FGD lasted from 50 to 70 minutes. In addition, three FGDs and six semi-structured interviews were carried out with the representatives of *aiyl okomotu* (local authorities), activists, and the most vulnerable population of Ak-Tuz, including pensioners and people with disabilities, as well as with project managers based in Bishkek, Brussels, and Osh who (had) worked in internationally funded programs in the field of mineral resource management and uranium remediation in Kyrgyzstan. A total of 18 respondents were interviewed.[5]

Empirical findings

This section presents the findings from the FGDs and semi-structured interviews that looked at people's perceptions of their capacities to deal with the social, economic, and ecological challenges affecting their communities. Because of the exploratory nature of our study, only a few, emblematic examples of coping, adaptive, and transformative forms of resilience are presented.

Coping capacities for resilience

During the interviews, the inhabitants of **Kadji-Sai** expressed widespread consensus that the main challenge facing their village is the lack of economic prospects. Families here are highly dependent on the limited state resources that are directly available to them, especially social benefits and pensions. Job opportunities are limited and gendered: for women, these include schools, kindergartens, and a recently opened tailoring business, and men can be taxi drivers or shift workers at the Kumtor gold mine located about 130 km from the village. Overall, unemployment is most prevalent during the so-called 'off-season' (in January and February). During the rest of the year, Kadji-Sai inhabitants cope by relying on self-sufficient agriculture and by cultivating fruits (e.g., apples, apricots, cherries for further sale) and crops in their own gardens. Because available land is limited (maximum 4 acres per family), such individual production is scarce, and only a small part of it is sold. During the tourist season, schoolgirls are employed as waitresses in cafes and as maids in guest houses, where they work 15 hours a day for a daily salary between 300 and 500 KGS.[6] Besides seasonal work in tourism and agriculture, keeping livestock (up to six animals) is practiced, although this activity is considered too expensive and unprofitable since the grassland is inadequate. In the winter months, people survive by doing so-called *zagotovki*—that is, preparing vegetables such as cucumbers, tomatoes, and cabbage and by salting and pickling them for the winter, both for their own consumption. Although migration has emerged as an adaptive strategy, this option is not available to all villagers owing to their lack of material resources and of relevant contacts outside the community. As highlighted by one of the FGD participants, 'You have to migrate to make good money, but you also need money to leave. So, this is a vicious cycle'.[7]

Mudflows are very frequent in Kadji-Sai. Only 10 minutes of heavy rain can result in mudslides that flow into the uranium tailing dump and through the village into Issyk-Kul Lake. Fortunately, remediation has been completed by the Russian company Rosatom. In such emergencies, the villagers react fast, relying on their bonding social capital. As reported by one of the respondents, 'When it starts to rain, we immediately begin to make ditches, follow the weather forecasts, prepare shovels, call up our neighbour'.[8] At the same time, the inhabitants of Kadji-Sai cope with the situation by stimulating local synergies, by coming together, collecting money, and cleaning the damaged channels that had been constructed during the Soviet era. As observed by one of the FGD participants, 'We restored the sump and treatment plant [and] installed and raised the level of pipes by ourselves'.[9]

Our second case study focused on **Orlovka and Ak-Tuz**, where the economic dimension was also perceived to be the most crucial challenge in the daily lives of most of the respondents. Interestingly, diversification of the capacities for resilience based on gender emerged in these locations. Indeed, in Orlovka, labour migration remains the most popular survival strategy for men as a way to cope with the socio-economic situation that resulted from the shutdown of the earlier city-forming enterprise, the Kyrgyz Chemical Metallurgical Plant. The men migrate either to other Kyrgyz cities (such as Tokmok or Bishkek) or to Moscow. In addition, young men also engage in short- and long-distance commuting to work for extractive industries that are offered by Altynken LCC and are available in Kara-Keche and Talas. In contrast, most of the women in Orlovka lack skills as well as a more diversified network of relations and are less mobile. As consequence, they tend to rely on the limited jobs available in their city. These jobs are poorly paid and are usually offered in the public sector, such as at two schools, kindergartens, a hospital, a music school, a house of culture (*Dom Kultury*), and the city library. In addition, there is a small-scale ski resort close to Orlovka that was built during the Soviet period to provide recreational facilities for the plant workers; however, in the 1990s it was privatised and divided among several owners. Today, the facility provides seasonal employment opportunities in service jobs mostly for female local residents.

Compared with the other locations, the socio-economic situation in Ak-Tuz is dramatically worse and offers fewer resources. According to the mayor, the local population consists mainly of elderly residents who rely largely on pensions and international humanitarian assistance, although such subsidies are not sufficient to meet their daily needs. Most of the locals keep livestock and gather and sell mushrooms at the end of summer and during the fall to prepare for winter, any proceeds going especially toward buying and delivering coal and hay. Some people economise by collecting industrial waste, the environmental legacy of Ak-Tuz's past as an industrial *monogorod*. More precisely, local residents dig up and sell scrap metal that can still be found in abandoned factories or at sites where industrial waste was stored and/or destroyed. This of course poses a crucial threat to their health and well-being.

In relation to this environmental risk, the ecological dimension across all research sites received less attention among our respondents. In particular, when they were asked about radiation and their personal and collective perceptions thereof, the respondents revealed a lack of scientific knowledge and a passive attitude when it came to dealing with this issue. In fact, most of the FGD participants in Orlovka insisted that there is no radiation, as articulated by one respondent: "We have a factory nearby, but we do not feel any radiation."[10] The majority of respondents in Ak-Tuz agreed that there are no dangers or any other radioactive risks and threats to public health from living in their village. Some of the study participants believed that they have been exposed to radiation for their entire lives and therefore have already adapted physically to living in such a contaminated environment as Ak-Tuz.

Adaptive capacities for resilience

As pointed out earlier with regard to the economic dimension, the decision to migrate – either to the capital Bishkek or to Russia – was reported to be one of the main survival strategies in both case studies: 'There are migrants in every second family. [...] They leave their children here with their grandparents and relatives'.[11] At the same time, migrants provide assistance to the village by, for example, giving financial support to Youth of Kadji-Sai, the only active local NGO. Through contacts with migrants living abroad (i.e., their bridging social capital), residents can gain access to tangible resources that are otherwise not available in the community, such as medicines, masks, and even food (which were brought in by migrants during the COVID pandemic for use by the villagers).

The strategies for dealing with water-related issues offer another good example of the adaptive capacities of the Kadji-Sai villagers. Overall, two main problems affect their water supply: its irregular availability and its contamination. Special containers were purchased, and filtering equipment was installed or constructed to purify drinking water that otherwise is swampy, full of debris and garbage, and, simply put, contaminated, 'since it is taken from a ditch also used by cattle'.[12] As a matter of fact, additional knowledge and skills are acquired by connecting with more knowledgeable actors and experts from outside the community, including the virtual space. As noted by one of the FGD participants, 'All residents do this. We look on the internet, if something needs to be bought, sellers also advise us'.[13]

Water distribution usually takes place according to a schedule that is not always respected. Indeed, people living near the watershed might arbitrarily interrupt the supply and redirect the flow of water based on their own needs. In such situations, the inhabitants of Kadji-Sai have learned an adaptive strategy based on reciprocal support, as highlighted by one FGD respondent: 'If someone does not have water, but guests visit him, he goes to negotiate with his neighbours so that they give him water for one hour'.[14] The villagers usually store water in buckets, barrels, and cisterns containing up to 500 litres, which allow the dirt to settle. Special pools that have been in use since the 1970s collect rainwater during the summer when the supply to households can be turned off to meet irrigation-related needs. In addition, water is collected in the neighbouring village of Bokonbayevo and brought by car to Kadji-Sai.

The inhabitants of Ak-Tuz provided another example of the ability of local residents to access external resources that are otherwise scarce in their village. Being confronted by a lack of a pharmacies and consequently of medicines, villagers give lists of their medical needs to local taxi drivers (including the *marshrutka* [shared taxi]) and pay them about 50 to 70 KGS to pick up and deliver their medicines.[15] Because of the geographical periphery of the village, residents apply the same strategy as a way to acquire other basic commodities – first and foremost, gas cylinders and food – from Kemin and Tokmok.

Transformative capacities for resilience

The FGDs revealed that, in sharp contrast to Kadji-Sai, the population of Orlovka has been successful in creating transformative capacity for resilience through the linking type of social capital. This has enabled their inclusion in decision-making processes, especially concerning social issues that affect their community. For example, at the time of our fieldwork, representatives from civil society and local authorities were engaged in organising a forthcoming community meeting by sending out invitations via WhatsApp groups. The topic for discussion was to be renovations at a city park that were scheduled for the spring of 2022. According to the FDG participants, such town meetings were frequent and were central not only for discussing common and highly important problems that affected the residents but also for preparing written statements on their behalf to be delivered to the mayor and the municipality. As highlighted by one of the most experienced female activists during the FGD, 'No project is implemented without the consent of residents reached at the community meeting'.[16] During the fieldwork, two groups of local activists were identified as being particularly active: the young residents from Orlovka and neighbouring villages of the Kemin district and senior female members of the so-called 'Women's Council'. These two groups are involved in organising both the regular community meetings attended by residents and other stakeholders (local authorities and sometimes representatives of a Chinese gold-mining enterprise) and the formal meetings that involve representatives from the international donor community. In this regard, it is worth noting that female activists have been developing technical competences and the skills required to successfully apply for financial assistance, and various small grant programs have been offered thanks to the support of international donors. In contrast to Kadji-Sai and Ak-Tuz, several public infrastructure objects (a hospital, two schools, sports facilities, and roads) have already been renovated in Orlovka, both with and without international financial support. Additional reparations of the urban infrastructure (including the main city park and a playground) are on the to-do list of local authorities.

With regard to the Ak-Tuz case study, we found that the potential for developing transformative capacities among the local community members was high, especially thanks to their contacts, engagement, and cooperation with local authorities and international donors (i.e., linking social capital). The newly appointed head of the village authorities arranges for small-scale community meetings to be held on a regular basis at which concrete problems can be discussed. A common topic at these meetings is the possibility to apply for international financial support for projects aimed at improving heating and water supplies, roads, and basic city infrastructure (depending on the season). In addition, a few local entrepreneurs have been running small-scale businesses funded by UNDP, allowing sewing and furniture shops to be opened and thus creating seven or eight job openings for the locals. As is the case in Orlovka, a group of people in Ak-Tuz who are familiar with the appropriate 'language' to be employed when approaching the donor community has recently come together to seek international financial support. As a matter of fact, in both cases we identified local activists and groups whose proactive attitude and entrepreneurial spirit facilitated their role as network brokers, linking residents and development agencies and assisting the former in accessing external resources.

Despite these positive trends and encouraging signs that local residents could develop transformative capacities for resilience, we noted some limitations, at least with regard to the environmental dimension. Indeed, the lack of knowledge concerning ecological challenges such as the air, water and soil contamination affecting these communities can be seen as a major obstacle when it comes to including *real people* in the decision-making processes concerning how to manage the uranium legacy sites.

Further Discussion

This study introduced a new analytical framework for assessing resilience in which each location investigated is considered a 'community of relations' (Korosteleva and Petrova, 2022, in this volume). This approach emphasises the centrality of the social ties and networks through which individuals can develop coping, adapting, and transforming capacities. By developing an *ad hoc* question-naire and by using the responses obtained through focus group discussions and semi-structured interviews with local residents, and stakeholders, we were able to connect these three capacities for resilience to three types of social capital (i.e., bonding, bridging, and linking). The data collected revealed how social capital theory can provide a useful lens for exploring a community's capacity to endure and transform itself – in this case, in the face of adversity within the context of post-Socialist transition (see also Pierobon 2019, and Pierobon 2021b).

Our analysis revealed several examples of how local residents engaged in forms of self-help to overcome immediate common threats by accessing resources that were directly available to them via family or neighbourhood ties. In many cases, the coping capacities described by the FGD participants relied on survival strat-egies and mechanisms found to be typical of other contemporary *monogorods* in transition. At the research sites we studied, the original town-forming enterprises that were created during the Soviet era were at first privatised several times and then finally shut down completely—an outcome perceived as a catastrophe for the individuals who stayed and that has led to large-scale crises. The FGDs also indicated that the level of adaptive capacities for resilience was low in Orlovka and Ak-Tuz but somewhat higher in Kadji-Sai. Interestingly, the (perceived) lack of support by local authorities and the international donor community has impelled the residents of Kadji-Sai to take steps to improve their livelihoods by applying what they have learned from past experiences and by connecting with individuals who have skills and resources that would otherwise be unavailable (i.e., bridging social capital). In addition, our data acknowledge that being in the geograph-ical periphery is a distinctive feature of the Soviet *monogorod* project and plays a crucial role in creating additional barriers to developing the bridging type of social capital, especially in the case of Ak-Tuz. Indeed, Orlovka and Kadji-Sai are taking advantage of their relative closeness to the main road leading to the cap-ital Bishkek, since this location allows them to improve their economic situations by intensifying contacts with other settlements, developing local tourism, and attracting visitors from Bishkek.

The ability to develop transformative capacities for resilience via the linking type of social capital in these communities is most hindered by the extent to which the residents distrust the local authorities (which was considerable in Kadji-Sai but relatively less so in Orlovka) as well as by a central government known for its widespread corruption. At the same time, the respondents frequently argued that these governmental institutions and structures have ignored them ('they completely forgot about us there'[17]) or consider these communities irrelevant, owing to their peripheral location geographically and lack of economic prospects. Another decisive factor that hinders the formation of ties between these targeted communities and international donors is the latter's tendency to interact and engage directly only with the local authorities' representatives instead of with *real people*. Paradoxically, it is the (perceived) lack of support by the local government and international donor community (i.e., linking social capital) that pushes people in Kadji-Sai to build adaptive capacities for resilience by connecting with people who are 'dissimilar' (i.e., bridging social capital). At the same time, however, the latter prefer to work directly with local authorities rather than with *real people*, making it difficult for the inhabitants of the target locations to build up their transformative capacities for resilience.

Conclusions

In line with the argument developed in the introduction to this volume, and by Korosteleva and Flockhart, who called for a double shift of focus at the level of the 'person'—from the 'external' to the 'internal' and from the 'global' to the 'local' (Korosteleva and Flockhart 2020, 155), our study explored how the concept of resilience has been perceived and experienced by the 'real people', in this case the residents of post-Soviet mono-industrial areas that are being affected by the uranium legacy and radiation in Kyrgyzstan. Despite the exploratory character of this study, our tentative findings serve multiple purposes and are relevant not only for monotowns but more generally for community development in the context of a post-Socialist transition. Overall, our data showed that the communities we selected for our case studies cannot be considered 'resilient', at least according to the descriptions of this phenomenon in the scholarly literature. As the previous sections illustrate, the inhabitants who continue to live at these sites developed a variety of coping or survival strategies in response to everyday life challenges. Fontaine and Schlumbohm (2000) pointed out that survival strategies are often seen as a complex phenomenon where the word 'survival' is not usually restricted to physical subsistence, but instead these strategies are 'always shaped by perception and self-perception, i.e., socially constructed' (ibid., 10). Importantly, the diversity of strategies that individuals or households might generate depends for the most part on the different types of uncertainty they face in their socio-economic, political, and cultural surroundings (ibid., 12). Even though some of the projects and initiatives funded by international donors are based on the concept of resilience, this term remains abstract for most of the people who reside in these targeted sites.

In the discourses and practices of international donors, resilience serves as a tool for promoting 'external engineering and security maintenance' (Korosteleva 2020, 5). In line with this logic, in the post-Socialist transition context with a focus on uranium-mining monotowns, projects geared toward stimulating community resilience are implemented based on the 'Western model' of uranium risk management – that is, the focus is mainly on raising awareness about environmental and health issues in the local population, which is accompanied by small grants to foster socio-economic development. However, our study shows the limitations of these top-down 'technical solutions', since the residents of the three research sites do not prioritise environmental and health issues over their daily struggle for survival. In fact, during the FGDs and interviews, local community members tended to emphasise the economic and social challenges they face and to use phrases such as 'try to survive' or 'make ends meet' ('араң күн көрүп жатабыз' in Kyrgyz or 'выживать' in Russian) when describing the current socio-economic situation in their locations. The international donor community could undeniably play a bigger role in strengthening the transformative capacities for resilience of target locations by tapping into local actors' full potential; however, for this to happen, they need to create bottom-up mechanisms and adopt an inside-out perspective by paying attention to local voices and visions for change. Support needs to be provided for projects that will 'enable[e] local communities and real people to actualize their potential in ways they [themselves] specify' (Korosteleva and Flockhart 2020, 159).

Our study sheds light on the community resilience concept as a possible way of achieving 'long-term thinking, living and governing' (Korosteleva and Petrova 2022 in this volume) that goes beyond the 'external' and temporary development interventions. The data we collected offered several examples both of community members who relied on forms of self-organisation without the need for governance or coordination from above and of communal support infrastructures that relied on the informal sharing of responsibilities typical of Central Eurasian countries (see Korosteleva and Petrova, 2022, 11, in this volume). Nonetheless, social capital theory also helps to point out the possible limitations when vulnerable groups are tasked with developing adaptive and transformative capacities for resilience, in that they the risk being relegated to a peripheral position within the so-called community of relations. Indeed, different social groups have different access to social capital because of their advantaged or disadvantaged structural positions and associated social networks. Our analysis of coping strategies confirmed that women's disadvantaged position limits their opportunities to form a variety of relationships outside the community (Blau 1977). Indeed, women are usually affiliated with smaller and less diverse female networks, their bonding ties being lower in a hierarchical position (Lin 2001). This lower status is even more entrenched in traditional societies of Central Asia, where women's networks are significantly shaped by their family and parental duties. Since such networks are homogeneous, they tend to reproduce these disadvantages with regard to women's access to external resources and power, thus contributing to gender inequalities. Therefore, the new understanding of resilience as an analytic of governance is far from being a panacea for bottom-up community transformation in the face of adversity and must still be handled critically by asking whose strengths and capacities are enhanced, whose local voices and needs are being heard, and whose visions are actually realised.

In addition, although a very rich social fabric of networks based on religion, traditions, and culture, is typical of societies in Kyrgyzstan, and in Central Asia more generally, these patterns of solidarity usually take place and manifest themselves through bonding ties and, less often, through bridging ties. Even though the Soviet experience could not dissolve this social fabric at the community level, its legacy is still evident in the communities' lack of trust in state institutions and authorities. Surely, this legacy still represents a major obstacle to making use of linking social capital and to developing transformative capacities for resilience, which would make it possible for community members to be a part of formal decision-making processes and thus influence policy on a regular basis.

Finally, our study contributes to this field by identifying new directions for future research. The findings presented here were based on the qualitative data generated over a short period of time. For future studies, we would recommend that longer-term fieldwork at the research sites be carried out along with participant observation techniques, thus affording valuable insights into the challenges posed by collective action and decision-making that involves several different groups. Finally, our study focused mostly on urban areas located in the country's northern part and are geographically close to the capital. Therefore, additional research at the legacy sites in the southern regions might provide valuable insights as to how further to build and strengthen community resilience in this fragile, post-Soviet region.

Acknowledgments

The authors would like to thank the funders and all the respondents who contributed to our research by participating in the focus-group discussions and semi-structured interviews. The authors are also grateful to Professor Elena Korosteleva, University of Warwick, and Irina Petrova UCL SSEES, for including their chapter in this monograph.

Notes

1. It is noteworthy that these activities also sought to increase acceptance of the recultivation and rehabilitation work on behalf of local communities by strengthening their participation in managing and governing the ULS (uranium legacy sites).
2. See more on this issue in Junussova and Beimisheva 2020; Nasritdinov 2015; Printsmann 2010.
3. According to an interview with a local mayor in January 2022.
4. Altynken LLC was established in 2006 and included a Kazakh investor and the state-owned gold-mining company Kyrgyzstan OJSC. However, in 2011, instead of the Kazakh company, a Chinese investor representing the 'Superb Pacific Limited Company' started negotiating a new licensing agreement that allocated 60% of the shares to Altynken and

40% to the state-owned company. This development has triggered a local resistance movement to prevent new mining projects in Orlovka. For more on the history of this issue, see Moldalieva and Heathershaw (2020) and Ocakli et al. (2021).

5. The initial set of key respondents at each site (schoolteachers, schoolchildren, activists, and officials) was extended via snowball sampling. Furthermore, respondents were selected based on the criteria indicating that they would have something to say on the chosen topics (see Burrows and Kendall 1997). FGDs were held at different sites in the target locations, such as schools, local government buildings, workplaces, and other public places. We obtained verbal informed consent from each person who participated in the FGDs and interviews for audio recordings and the confidential and exclusive use of data for research purposes. Transcripts of the interviews and the FGDs were analysed using a systematic data coding scheme based on three main categories (social, economic, and ecological challenges) and the six subcategories: coping, adaptive, and transformative capacities for resilience, and bonding, bridging, and linking social capital. In addition, the observational notes taken during the fieldwork were used to better contextualise the FGD and interview data.

6. Equivalent to 3.49 to 5.81 USD (April 2022).
7. FGD with women in Kadji-Sai, November 2021.
8. FGD with women in Kadji-Sai, November 2021.
9. FGD with men in Kadji-Sai, November 2021.
10. FGD with the female activists in Orlovka, December 2021.
11. FGD with activists in Kadji-Sai, November 2021.
12. FGD with men in Kadji-Sai, November 2021.
13. FGD with men in Kadji-Sai, November 2021.
14. FGD with women in Kadji-Sai, November, 2021.
15. Equivalent to 0.50 to 0.70 cents USD (March 2022).
16. FGD with the female activists in Orlovka, December 2021.
17. Interview with a former head of local authorities in Ak-Tuz, January 2022.

Disclosure statement

No potential conflict of interest was reported by the authors.

Funding

This research project took place within the Postdoctoral Fellowship Programme entitled 'Institutional Change and Social Practice. Research on the Political System, the Economy and Society in Central Asia and the Caucasus', funded by the Volkswagen Foundation in Germany.

ORCID

Chiara Pierobon ⓘ http://orcid.org/0000–0002–1398–4665

Zarina Adambussinova ⓘ http://orcid.org/0000-0003-1014-8378

References

Aldrich, D. P. (2012) *Building resilience: Social capital in post-disaster recovery*, (Chicago,: University of Chicago Press).

Bernier, Q. and Meinzen-Dick, R. (2014) 'Networks for resilience. The role of social capital', *Food Policy Report*, accessed at: http://dx.doi.org/10.2499/9780896295674 (3 March 2022).

Blau, P. M. (1977) *Inequality and Heterogeneity: A Primitive Theory of Social Structure.* (New York: Free Press).

Bogdetsky, V., Ibraev K., and Abdyrakhmanova J. (2005) *Mining industry as a source of economic growth in Kyrgyzstan*. Bishkek, Kyrgyzstan: PIU of World Bank IDF Grant for

Building Capacity in Governance and Revenues Streams Management for Mining and Natural Resources.

Bogdetsky, V. and Novikov, V. (2012) *Mining, Development and Environment in Central Asia: Toolkit Companion with Case Studies*, Zoï Environment Network, University of Eastern Finland, Gaia Group Oy, Joensuu, Finland.

Doolot, A. and Heathershaw, J. (2015). 'State as resource, mediator and performer: understanding of local and global politics of gold mining in Kyrgyzstan', *Central Asian Survey*, 34, 93–109.

European Bank for Reconstruction and Development (EBRD). n/a. The Environmental remediation account for Central Asia (ERA), accessed at: www.ebrd.com/what-we-do/sectors-and-topics/nuclear-safety/era.html (17 October 2021).

Ejdus, F. and Juncos, A. E. (2018) 'Reclaiming the local in EU peacebuilding: Effectiveness, ownership, and resistance', *Contemporary Security Policy*, 39:1, 4–27.

European Union (2016) *Shared vision, common action: A stronger Europe, a global strategy for the European Union's foreign and security policy.* (Brussels: European Union).

Fontaine, C. and Schlumbohm, J. (2000) 'Household strategies for survival: An introduction', *International Review of Social History*, 45, 1–17.

Granovetter, M. (1983) 'The strength of weak ties: a network theory revisited', *Sociological Theory*, 1, 201–233.

Hawkins, R. and Maurer, K. (2010) 'Bonding, bridging and linking: How social capital operate in New Orleans following Hurricane Katrina', *British Journal of Social Work*, 40, 1777–1793.

Humphrey, P. and Sevcik, M. (2009) 'Uranium Tailings in Central Asia: The Case of the Kyrgyz Republic', accessed at: www.nti.org/analysis/articles/uranium-tailings-kyrgyz-republic/ (12 December 2021).

Jacobs, E. and Hofman, I. (2019) 'Aid, social capital, and local collective action: attitudes towards community-based health funds and village organizations in Rushan', Tajikistan, *Community Development Journal*, 55: 3, 399–418.

Junussova, M. and Beimisheva, A. (2020) 'Monotowns of Kazakhstan: Development challenges and opportunities'. In: Koulouri A., Mouraviev N. (eds.): *Kazakhstan's developmental journey: entrenched paradigms, achievements, and the challenge of global competitiveness.* (Singapore: Palgrave Macmillan), 211–247.

Keck, M. and Sakdapolrak, P. (2013) 'What is social resilience? Lessons learned and ways forward', *Erkunde*, 67: 1, 5–19.

Korosteleva, E. (2020) 'Reclaiming resilience back: A local turn in EU external governance', *Contemporary Security Policy*, 41: 2, 241–262.

Korosteleva, E. and T. Flockhart. (2020) 'Resilience in EU and international institutions: Redefining local ownership in a new global governance agenda', *Contemporary Security Policy*, 41: 2, 153–175.

Korosteleva, E. and Petrova, I. (2022) 'What makes communities resilient in times of complexity and change?', *Cambridge Review of International Affairs*, DOI: 10.1080/09557571.2021.2024145

Kwok, A. H., Becker, J., Paton, D., Hudson-Doyle, E. and Johnston, D. (2019) 'Stakeholders' Perspectives of Social Capital in Informing the Development of Neighbourhood-Based Disaster Resilience Measurements', *Journal of Applied Social Science*, 13.

Lin, N. (2001) 'Inequality in social capital'. *Contemporary Sociology*, 29: 6, 788.

Magis, K. (2010) 'Community Resilience: An Indicator of Social Sustainability'. *Society & Natural Resources*, 23: 5, 401–416.

Moldalieva, J. and Heathershaw, J. (2020) 'Playing the "Game" of Transparency and Accountability: Non-elite Politics in Kyrgyzstan's Natural Resource Governance', *Post-Soviet Affairs*, 36:2, 171–187.

Morgan-Trimmer, S. (2013) '"It's who you know": community empowerment through network brokers', *Community Development Journal*, 49: 3, 458–472.

Nasritdinov, E. (2015). 'Deurbanization: the ruins of the Soviet modernism in mining towns of Kyrgyzstan'. *CABAR.Asia* (available in July 2018).

Ocakli, B., Krueger, T. and Niewöhner, J. (2020) 'Shades of conflict in Kyrgyzstan: National actor perceptions and behaviour in mining', *International Journal of the Commons*, 14: 1, 191–207.

Ocakli, B., Krueger T., Janssen M. A., Kassymov U. (2021) 'Taking the discourse seriously: Rational self-interest and resistance to mining in Kyrgyzstan', *Ecologic Economics*, 189.

Onyx J. and Bullen P. (2000) 'Measuring Social Capital in Five Communities', *The Journal of Applied Behavioural Science*, 36:1, 23–42.

OSCE (n/a) 'Strengthening stakeholder participation for uranium legacy remediation in Kyrgyzstan', accessed at: www.osce.org/files/f/documents/c/b/285021.pdf (10 September 2021).

Petrova, I. & Delcour, L. (2020) 'From principle to practice? The resilience–local ownership nexus in the EU Eastern Partnership policy', *Contemporary Security Policy*, 41:2, 336–360.

Pierobon, C. (2019) 'EU's support to civil society in Kazakhstan: A pilot evaluation of the social capital generated', *Journal of Evaluation*, 25:2, 207–223.

Pierobon, C. (2021a) 'EU Efforts to Prevent Violent Extremism (PVE) by Engaging Civil Society in Kyrgyzstan', *Central Asian Affairs*, 8, 150–174.

Pierobon, C. (2021b) 'Social Capital's "Missing Dimension" in Sustainable Development Programs: Evidence from EU Support of Civil Society in Kazakhstan', in C. Pierobon, N. Becker and S. Schlegel, eds, *Central Asia After Three Decades of Independence. Politics and Societies Between Stability and Transformation*, Nomos, Baden-Baden, Germany, 125–150.

Printsmann, A. (2010) 'Public and private shaping of Soviet mining city: Contested history?' *European Countryside*, 3, 132–150.

Reiserer, A. (2021) 'Uranium Legacy in Central Asia: Significant progress, but funding gap remains', accessed at: www.ebrd.com/news/2021/uranium-legacy-in-central-asia-significant-progress-but-funding-gap-remains.htm (10 September 2021).

Steiner, A. and Markantoni, M. (2013) 'Unpacking community resilience through Capacity for Change', *Community Development Journal*, 49: 3, 407–425.

UNDP (2019) 'Stakeholder Engagement for Uranium Legacy Remediation in Kyrgyzstan'. Phase II - Regional Project Document, accessed at: https://info.undp.org/docs/pdc/Documents/SVK/Project%20Document%20-%20Uranium%20II.pdf (10 September 2021).

Vertigans, S. and Gibson, N. (2020) 'Resilience and social cohesion through the lens of residents in a Kenyan informal settlement', *Community Development Journal*, 55:4, 624–644.

Woolcock, M. and Narayan, D. (2000) 'Social capital: Implications for development theory, research and policy', *World Bank Research Observer*, 15:2, 225–249.

Conclusion
Locating Central Eurasia's inherent resilience

Prajakti Kalra ⓘ

Abstract *This article aims to contextualise the inherent resilience of Central Eurasian states through the exploration of their particular history. The main purpose is to ground the ideas of resilience and capacity building in the context of the geography and ecology of Central Eurasia thus confronting the current views of the need for making these communities resilient by borrowing European, Western or global 'best practice' in order to achieve stability and development. This paper offers an overview of the history of the region to bring into focus the 'local' Central Eurasian milieu. The sophisticated tapestry of understanding, action and strategies developed over centuries has made this region resilient in the face of unpredictability caused by natural and manmade events. This paper seeks to locate how the region has consistently overcome obstacles in its long history of inhabiting a disparate space. We apply the term intercalation here to describe the emergence of a collective identity from strongly interacting ingredients that represents the inherent resilience of the region. Consequently, the focus is on the ways in which communities within the region connect, cooperate and build nodes of interaction to achieve prosperity and development.*

Locating Central Eurasia

Central Eurasia is home to some of the oldest and most diverse civilisations in the world. Geographically it stretches from the Caspian Sea to Western China and includes modern day Central Asia, Caucasus and the Danube Delta. The term Central Eurasia is used here to accommodate the aspects of the region that go well beyond geographical borders of nation states today and refers to the expansiveness of the region that encompasses Central Asia and Inner Asia (Levi 2020, xiv). The actual terminology applied to the region has been a subject of scholarly debates and varies according to the time period (see Khazanov and Crookenden 1986; Frank 1992; Cowan 2007). Historically, it has accommodated an admixture of nomadic, semi-sedentary and sedentary peoples of Turkic and Iranian descent who have inhabited the large expanse of Eurasia for centuries. From the earliest migrations to cataclysmic natural events like earthquakes and climate change, monumental shifts have forged these communities that are at once the oldest and newest conglomerations of peoples. Located as Central Eurasia is on the historic Silk Road(s), the region has been the famed contact zone for diverse groups for at least two thousand years. Historically, this space has experienced a high level of mobility and interaction as a result of nomadic and semi sedentary social and cultural systems. It has served as the central node of significant connectivity across the Eurasian space, especially from the thirteenth century.

Post-1991 saw a monumental shift in the region with the dissolution of the Soviet Union. Central Eurasia for the purposes of this study includes the former Soviet republics of Kazakhstan, Kyrgyz Republic, Tajikistan, Turkmenistan and Uzbekistan. Since independence, these countries have been variously subsumed under categories of post-Soviet, -Socialist, -Communist among others and have been included in the ranks of underdeveloped countries. This has meant specific things in the context of the development discourse with international organisations and European and Western developed countries offering formulaic solutions to perceived Soviet inefficiencies. While this was genuinely welcomed by the Central Eurasian countries in the early 1990s, the last decade has seen remarkable levels of scepticism and questioning of these 'one size fits all' solutions (Kalra and Saxena 2021). This has led to a shift to the 'local' and has initiated processes ranging from challenging neo-liberal notions of politics, economics and society to the realisation for the need to embrace differences. With this in mind, this paper focuses on locating Central Eurasia in historical context with a view to describing its inherent resilience through the exploration of intercalation (Weller et al. 2005).

The paper is interdisciplinary and will inform the disciplines of history, international relations, political science and area studies. The main contribution will be to highlight the role cooperation has played in Eurasia and that the architects of that cooperation have been Central Eurasian actors in history. The historical backdrop serves to enrich the discourse on Central Eurasia and is offered as a tool to reimagine the communities that have simultaneously inhabited the centre and the periphery. There is an attempt at synthesising historical information with, broadly speaking, sociological and political processes at play in the modern era. The categories of analysis are historical and are used to explain trends in behaviour over an extended period of time. This particular rendition of events speaks to the shift towards decolonising disciplines. International relations studies have attracted criticism for an over-reliance on European and Eurocentric (Westphalian) notions of nation-states to understand world politics (Megoran and Sharapova 2013). This paper aspires to speak to the importance of decolonising international relations and go further by calling for de-sedentarising historical discourses which overwhelmingly focus on qualities of sedentary societies as a benchmark for civilisation (Rouse 2020). We attempt to locate nomads and steppe polity within Central Eurasian history and, as such, use this as an initial step towards an aspiration to call for a change in modes of thinking and analysis. This is done primarily through the application of the concept of intercalation that highlights Central Eurasia as a zone of intense interaction between different polities, peoples and ideas that serves to describe these societies as resilient in time and space.

We begin by providing a concise understanding of resilience in an increasingly uncertain world and introduce the concept of intercalation to explain Central Eurasian (actors, states and communities) resilience. In order to do so, we use theories of international relations to locate Central Eurasia in the post-1991 world order. Next, we offer a summary of the historical trajectory of the region. Following that we identify two categories of analysis, namely Eurasia and the Empire, to contextualise Central Eurasian history for the purposes of norm-making and behaviour. This is followed by giving the context of Eurasia's geography and exigencies of medieval empires to explain how connectedness lies at the very heart of governance and development in the region throughout history. We use the example of traders in the 13th century among others to illustrate how Central Eurasians translated different cultures and found interconnected spaces that led to greater prosperity. The next section traces regional integration in the modern era to express the region's continued desire for cooperation and functional connectivity. In this section, we juxtapose regional integration processes with the social constructivist theory in international relations. Finally, the conclusion highlights the need for understanding Central Eurasia within its particular historical context, highlighting intercalation processes, to understand the inherent resilience of these communities.

Contextualising resilience

The concept of resilience and good governance are increasingly used as a tool by international organisations like the European Union, World Bank, IMF and OECD, among others, to teach good practice in governance to affect capacity and build resilient communities in developing countries. In this context, resilience is defined as the capacity to recover quickly from difficulties and the ability of an object to spring back into shape especially in a VUCA (vulnerable, uncertain, complex and ambiguous) world (Bennis and Nannus 1986). Scholars have proposed resilience-thinking to explain a variety of issues from the management of natural resources (Walker et al. 2006, xiii) to the behaviour of cooperative organisations like the Non-Aligned Movement (Vieira 2016). Vieira, for instance, used the idea of a shared identity as the basis of continued resilience, and scholars studying international relations have expanded resilience to study a number of areas including economic reforms, civil society and governance globally.

The concept has attracted criticism in recent times as it is increasingly seen as a way to explain strategies of governance especially in a neoliberal context. Scholars have critiqued this western-oriented export of 'good governance' that is seen as an instrument to impose 'western' thinking and governance on developing states (Chandler 2010, 2014). Furthermore, challenges posed by poverty, disease, environment and climate change have created complexity in the world that is calling into question the application of resilience-thinking in the way international organisations are using it. More recently, there have been calls to shift the focus to the 'local' (Chandler 2014; Korosteleva and Flockhart 2020) that is better able to instil sustainability and responsiveness in governance. In this paper and the rest of the special issue, there is an attempt to explain the resilience of the region of Central Eurasia as a 'quality of complex adaptive system' and 'a new analytic of governance' by bringing the 'local' into focus. This is done specifically to propose the building blocks of a 'good life' in the context of the local environment (Korosteleva and Petrova 2022). An associated concept is that of peoplehood which, for the purposes of this collected volume, is primarily community driven and highlights the Central Eurasian historical perspective.

International theory often prioritises state actors in determining agency, process and social structures. It is important to point out that state actors represent communities and peoples on the ground and are an outcome of the local milieu that determines behaviour and practice. The local context of Central Eurasia needs cognisance to begin the process of understanding the region, including the ability of its states and its people, to inform good governance, capacity building and indeed the pursuit of a 'good life' whether from within or with help from outside. As Mearsheimer puts it, 'there are no universal truths regarding what constitutes a good life' and disagreements regarding what constitutes a good life can be profound (Mearsheimer 2018, Chapter 2: 7). To help locate the local context we turn to Holling who ascribes three properties to an adaptive complex system: wealth (inherent potential), controllability (internal connectedness) and adaptive capacity of any ecosystem, individual and culture. Any social or economic system can derive inherent potential from skills, network of relationships or mutual trust (Holling 2001, 394). Building on this, the long arc of history is used as the basis of Central Eurasian identity-formation, community, nation-state and region that taken together represent the process of intercalation. The main category of analysis is historical and we use extant sources that span steppe and sedentary interactions in time. The use of secondary sources helps provide a comprehensive picture that is used mainly to illustrate elements that emphasise cooperation in Central Eurasian communities through its history. The making of these communities in empires, Khanates and the Soviet Union are the necessary frames through which state and community behaviour should be understood.

Intercalation processes in Central Eurasia

The process of intercalation captures the essence of how these societies found themselves adjacent to numerous others and yet continued to retain their inherent properties to respond efficiently to increasingly complex environments. In Central Eurasia, a collective identity emerged from those strongly interacting elements that responded to VUCA worlds of the past through locating connected spaces. Those interactions referred to as 'repositories of action' (Neumann & Wigen 2018) represented Turko-Mongol and Turko-Persian Central Eurasia (Canfield 1991), as examples of intercalation processes, discussed below. In addition, the region is prone to earthquakes that represent immense unpredictability. These have shaped the ecology, society and economy of the region (Jackson, Earthquakes Without Frontiers Project, 2012–2018). The combination of manmade and natural events, often cataclysmic, left indelible marks on these societies to give rise to resilient communities. We use instrumental and civilisational (western and non-western) understandings of international relations and perceive the role played by Central Eurasia in macro level regional organisations in the modern era as a way to reinforce the desire and need for connectedness. Regional organisations are an example of the process of intercalation in motion since they represent national interests of Central Eurasian countries in a regional context in the present day. Subsequently, the focus is on the tendency of Central Eurasia (peoples and states) to connect as a fundamental response to unpredictability in the world around them. The resilience that emerged out of this shared geography saw these societies continuously trying to connect and find partnerships in response to external challenges and transformations through history. Based on this repeated connectedness we ascribe a collective regional identity to Central Eurasia that is more in tune with cooperative world orders (Wendt 1992).

Disruptions and reconfiguration

The uncertain, unpredictable (Bousquet and Curtis 2011) and complex world (Kavalski 2007) of today is a far cry from what was predicted in the early 1990s. The dissolution of the Soviet Union was seen as the triumph of the American led rules-based world order that was born out of World War 2 and the Cold War era and marked the end of history (Fukuyama 1992). Ikenberry (2005) described the new world order as the age of an American led global order loosely based on rules of open markets, security alliances, multilateral cooperation and democratic community. The end of the 1990s saw the creation of an unprecedented political system with America at the helm of a trans-oceanic alliance with Europe and other democratic states that were tied together by the market, institutions and security partnerships (Ikenberry 2005, 133–39). Central Eurasian countries arrived into this world system that wanted to fashion images of them in the post-Soviet space but one that was not prepared to accept the diversity that they brought with them.

The main international relations schools: realism, liberalism and social constructivism view cooperation and cooperative orders in different ways. For the realists, cooperation is not part of the nature of state and thus impossible; while liberals believe processes (institutions) can help evolve cooperative behaviours over time. For realists, liberals, neo-realists and neo-liberals the possibility of cooperation does not exist or exists only under specific conditions of international institutions (Keohane 2005). While the realist perspective is unable to engage with the new security dilemmas at hand, such as issues of climate change, natural resource management, poverty, hunger, financial upheavals and indeed disease that pose a far more real threat to modern day nation-states than war; the neoliberal dream and the promise of institutional theory have also fallen short in addressing these problems (Mearsheimer 2018). The world has continued on a trajectory that appears to have increased complexity (Holling 2001) and brought forth challenges that cause yet more dissonance and disruption, and pit civilisations against each other (Huntington 2011). Specifically, the imposition of western ideologies and concepts has reintroduced colonial agendas in understanding Central Eurasia (Dadabaev and Heathershaw 2020).

Social constructivists, especially Alexander Wendt, consider anarchy an outcome of what states make of it thus making collective action and cooperation possible (Wendt 1992, 395, 402). They believe that the domestic society creates the raw materials necessary for a collective regional identity that views cooperation as possible and even desirable (Wendt 1992, 403; Milner 1992). In this context, the very notion of identity formation that lies at the heart of social constructivism when applied to this region rests in the ability of Central Eurasian nations to grab the reins of a Eurasian narrative and imbue it with the lessons of their lived experience that created a collective stable identity, irrespective of fixed territoriality (Wendt 1992). The analytical frames to understand Central Eurasian societies are located in the historical trajectory of the Eurasian space (geography and ecology) that continues to select connectivity over disconnectedness. The nature of the peoples of Central Eurasia as interlocutors in this vast space instils characteristics of cooperation in these societies. In other words, in being connected the region finds itself individually and collectively prosperous, making connectedness the heart of this particular discourse.

Central Eurasia in the making

The peoples of Central Eurasia represent nomadic, semi-sedentary and sedentary civilisations in history. They have their own languages and cultures which are a product of nomadic and sedentary traditions in the steppe, oases cities and urban centres. Warrior societies like that of the Scythians, Xiongnu (Huns), Kok Turks (sixth century), Uyghurs (eighth century) and Mongols (thirteenth century) were categorically different from their more sedentary neighbours in China. Beginning with the Xiongnu (third century BC), there is clear evidence of nomads with their own political and social make-up. The evolution of nomadic society with its own hierarchical structure, leadership and economy flourished on the Eurasian steppe (Sneath 2007). Archaeological evidence, placed alongside Chinese textual analysis, paints a vivid picture of China's contacts with the Eurasian steppe peoples as far back as the Bronze Age. Interactions were often militaristic, diplomatic and trade related (Di Cosmo 1999). A stable network of exchanges included horses, cattle and other pastoral products imported by China, while there is also evidence of jewellery, weapons and other objects from China in tombs in steppe regions (Di Cosmo 1999, 957). These interactions led to shared understandings in a space that was constantly in contact. Accordingly, Central Eurasian societies displayed processes of intercalation in how they found ways to coexist while still maintaining distinct communities to display diversity in civilisational cultures (nomadic-sedentary and pastoral-agricultural) that 'echoed through later regional cultures and into the socio-politics of the present day' (Rouse 2020, 399–400).

Other than China, pre-Islamic Persia also had interest in Eurasia and there was expansion into Central Eurasia by the Achaemenid (4–5 century BC) and Sassanian (3–7 century AD) empires. They left a lasting Persianate imprint on society. With the Arab conquest of the eighth century and the Battle of Talas, Chinese expansion came to a close and the Islamic period of Central Eurasia took hold. The Caliphate brought with it Arab and Islamicised Persianate influences that continued well into the Samanid period (819–1004 AD). These features of pre-Islamic Persia and the Islamicate world (Arab and Persian) mingled with the nomadic entities, especially under the ruling houses of the Qarakhanid (tenth century), Ghaznavids (tenth century) and Seljuqs, and reached a climax under the Chingissid Mongols (thirteenth century). Subsequent Khanates (Timurid, Shaibanid and Kazakh Khanates) across Eurasia continued to pay homage to Chinggis Khan and the Mongol Empire as the source of political legitimacy, rule and governance well into the eighteenth-nineteenth centuries. From the thirteenth to the nineteenth centuries, society was organised around a ruling class made up of elites of Iranian and Turkic descent, the language of court (government, art and literature) was

Persian, and Arabic served as the formal language of religion and education (Robinson 1997, 2010). This multi-civilisational and multi-ethnic society should be viewed as an illustration of the processes of intercalation wherein Central Eurasian society had strong influences from at least three different racial and cultural groups that were accommodated and represented in individual and collective practice in the region. These elements were visible not only in Central Eurasian empires but were evident further afield in India and Iran as well. The region-wide shared dynastic tradition included rule driven by power (not religion), inclusive religious policies, prominent role of women in society (ruling and patronage), and architecture and art that spoke to the grandeur and power of the ruling elites (Robinson 2007). Furthermore, the Bukharan, Khivan, Kokand and Kazakh Khanates in Central Eurasia continued to operate within this same cultural and political realm, exhibiting resilience in subsequent time periods. As mentioned earlier, Canfield (1991) described Central Eurasia as Turko-Mongol and Turko-Persia in character, which can be referred to as the process of intercalation in which strongly interacting disparate civilisations and societies of Eurasia were able to flourish together and still maintain their innate features. This is not to say they did not change, but rather that they displayed adaptive capacity thinking by fashioning new institutions in response to challenges. For instance, nomadic and sedentary institutions were able to operate in parallel in Central Eurasia after it became part of the Mongol Empire in the thirteenth century that, in turn, made it an empire of even greater purport (Kalra 2020a).

The advent of the Tsarist Russian Empire at the beginning of the eighteenth century began the process of European colonisation that reached a culmination with the Directorate of Turkestan in the nineteenth century that included nearly all of present-day Central Eurasia (Khalid 2000). The Bolshevik revolution in 1917 and the eventual shift of power to the Soviet Union saw the National Delimitation process that became the foundation stone for the now sovereign nation states of Central Eurasia (See Hirsch 2005; Martin 2011; Roy 2000). This continuum of connectivity (physical, political and economic) was significantly impacted in 1991 with the first major disruption in the region when flexible borders became international borders characterised by military posts and barbed wires. Until 1991, nomadic empires, Khanates, European empires and even the Soviet Union had maintained a mostly connected Central Eurasian landscape. It was only with the advent of independent nation-states post-1991 that there was a fracture in the physical connectivity in the region that impacted mobility in all spheres. However, soon after 1991 we saw a number of cooperative organisations: economic and political, established in the region to respond to the disruption. These initiatives driven by multiple players in Eurasia including Kazakhstan, Uzbekistan, Russia, China, Japan and even Turkey indicate a region-wide understanding and desire for connectivity (Murashkin 2018). These include, but are not limited to, the Commonwealth of Independent States (CIS), Collective Security Treaty Organisation (CSTO), Shanghai Five and the Shanghai Cooperation Organisation (SCO), Eurasian Economic Community (EurAsEc), Central Asian Union (CAU), Free Trade Zone, Central Asia Plus Japan, Black Sea Forum for Partnership and Dialogue, Black Sea Economic Cooperation, Caspian Summits and the Central Asia Regional Economic Cooperation (CAREC) Program. The Eurasian Economic

Union (EAEU) and China's Belt and Road Initiative (BRI) are the latest examples of such groupings. These cooperative organisations serve a number of security and economic purposes and are platforms to build and maintain norms in the wider Eurasian region (Kalra and Saxena 2007; 2015).

Central Eurasian categories of analysis

Geography in large part has dictated the position in which these communities and states find themselves in, that is the heart of Eurasia. Any reading of Central Eurasian history places it well in the midst of facing multiple challenges (invaders, earthquakes and climate change) over centuries. The history of the region introduces meaning in the behaviour of these so called 'in-between' nations that echoes the past and yet serves as rhythms for the future (Saxena 2019, Preface). The two categories of analysis of Eurasia and Empire together explain the emergence of an identity that evolved from a connected Eurasian landscape and makes the region inherently resilient in the face of challenges. These categories also serve to illustrate the making of Central Eurasian societies via the processes of intercalation, reminiscent of multiple understandings and belongings established over time through strongly interacting groups, cultures and peoples.

Eurasia as a unit of analysis

Eurasia, straddling Asia and Europe, covers roughly 36.2% of the Earth's surface. It includes Central Eurasian countries, namely Kazakhstan, Kyrgyz Republic, Tajikistan, Turkmenistan and Uzbekistan among post-Soviet states and parts of neighbouring countries. Land connectivity is, in many cases, the only way that these countries can get their goods and products out into the global market as they are landlocked and, in the instance of Uzbekistan, doubly landlocked. It is important to point out that the idea of Eurasia is separate from Russocentric (Kotkin 2007), Kazakhstani (Mostafa 2013), Turkish (Akçali and Perinçek 2009) or any other such Eurasianism. The basis for talking about Eurasia here is not to call on a political or cultural understanding of Eurasia, but rather offer a geographical and historical underpinning of the region of Central Eurasia. It is clear that over a period of time, medieval empires (nomadic and European) and the Soviet experience established a repeated reciprocal typification in the region that has led to stable conceptions of self and the other for Central Eurasian societies (Wendt 1992, 405). The raw materials of understanding the self and the other were already in place in 1991 and can help further the understanding of peaceful change in the region (Dadabaev and Heathershaw 2020). Not only do the communities in the region have a long history of interacting with each other as discussed above but they also have a specific language to understand each other. What lay underneath the Soviet division of Central Eurasia was a Turko-Persianate-Islamicate-Mongol world order, mentioned above, that created mutually intelligible identities and modes of interaction (Canfield 1991). Canfield used a Greek term, Ecumene, to denote Central Eurasia as a 'historically perpetuated complex of meaningful forms' and a 'world of shared understandings' (Canfield 1991, xiii). Together these aspects of society represent the basis of behaviour and practice even today highlighting processes of intercalation in Central Eurasia.

Eurasia, articulating a specific geographic zone, can offer a more inclusive category of understanding. It explains cooperation and collaboration as the backbone of community and state in the region. The Eurasian ecology, encompassing the steppe, Tundra, wet lowlands, forests and deserts, gave rise to diverse peoples that have a particular relationship with their environment and each other (Kotkin 2007, 499). The diverse polities of Eurasia exerted pressure on each other throughout history that led to modes of interactions that connected the known world in multiple ways (Abu-Lughod 1989). However, these became concealed by Western imperialism of the nineteenth-twentieth centuries (Robinson 2007). Many of these interactions date further back than any European contact with either China or Central Eurasia and continue well into the nineteenth century. In order to capture the motivations of these people, it is essential to follow their particular historical path and avoid the mistakes of giving meaning to their actions that are alien to them. Very much like the sedentary narratives that belie nomadic-sedentary civilisational contact and action, European understandings ignore the importance of Eurasian connectedness, and the post-Cold war American led world order sees Central Eurasia through the prism of Russia, and now China. Taken together, they give a distorted view of the region and perpetuate misunderstandings. The other papers in this volume touch on the different philosophical, intellectual (Chandler) and religious underpinnings from Islam, Sufism (Nurulla-Khojaeva; Shahidi), lived community (mahalla) and neighbourhood (Dadabaev) that maintained these conditions of connectedness. From the smallest atom of the self to the largest community, and the nation-state, the wider Eurasian region has its own lexicon of institutions, connected living spaces and understanding of the immediate (neighbours) and the distant (international) other.

Central Eurasian history removed from the diluted visions of either purely sedentary and/or colonial worldviews can instil meaning in the actions and practices of actors in the region today. This has implications for both external and internal practice. Here external implies relations with the western world: US and Europe but also with Outer Eurasia (Russia and China), while internal implies interactions between Central Eurasian states and the communities that lie within. It is precisely the historic interactions between nomadic-sedentary civilisations in medieval empires that need to be taken into account before giving meaning to state or community action and behaviour (Wendt 1992). In other words, Europe and Asia view each other in specific ways based on centuries of interactions; however, nation-states like that of Central Eurasia are too often unable to participate in the mirror-image theory of identity formation directly (Wendt 1992, 407). They have appeared in scholarship and politics through proxy in time and space: China, pre-Islamic Persia, Islamic world, Russia and the Soviet Union, and now modern Russia and China. This identity formation imbues characteristics in Central Eurasian state actors and communities that do not always belong to them but are borrowed or expanded from their seemingly more significant and powerful neighbours in history and the modern era, respectively. However, within Eurasia there are heuristic tools of empire (Chinggisid Mongol for instance) and Inner Eurasia (Christian 1994) that provide meaning and practice for identity formation and behaviour significant for the present. The rules of engagement have a basis in historical interactions and behaviours that carry meanings that have been at play over a long time and even include partnerships between pastoralists and agriculturalists in antiquity (Rouse 2020, 400). In terms of governance, both sustainable and reliable, it is thus essential to view the features that have helped create communities of peoples in the region. These characteristics include a tendency toward cooperation, collaboration and co-optation, borne out of geographical and ecological exigencies.

Empires as a unit of analysis

Historians have long questioned the significance of using the nation-state as a unit of analysis because it dictates the recording of events in the twenty-first century in very specific and often alien ways for a large swathe of humanity. David Christian's regional analysis of Inner Eurasia juxtaposed with Outer Eurasia is a worthy attempt at shifting the lens from the nation-state in the region (Christian 1994). In his rendering of events, the focus shifts to the nomad and the steppe and lends a political coherence to the region that rests in geography and ecology to give rise to certain types of empires (Christian 1994, 175). His analysis bridges the history of Central Eurasia all the way from antiquity to the Soviet Union and, counter-intuitively, brings the impact of Central Eurasia on Outer Eurasia (China and Russia) to the fore. Furthermore, this shared legacy, explained primarily through the unit of empire, is confirmed by Robinson's work on Mughal, Safavid and Ottoman empires in wider Eurasia (Robinson 1997). Specifically, these empires accommodated multiple belongings and understandings: cultural, political and economic, and shared connective knowledge systems (formal learning and esoteric understandings), commercial organisation and techniques of trade (Robinson 1997, 151). These multi-civilisational, multi-ethnic and polyglot empires of the steppe signify monumental shifts in world history but have been superseded by European hegemonic discourses (Goody 2006). Significantly, the interplay between nomadic and sedentary institutions within these particular empires created conditions that nourished adaptability in governance, openness to ideas, and a mixed economy that accommodated agriculture, pastoralism and trade across the large expanse of Eurasia (Kalra 2018). Taken together, the exigencies (climactic, civilisational, religious, cultural and political) the region was exposed to through history help explain the adaptability of communities within Central Eurasia to withstand shocks and challenges by utilising connected spaces where disparate groups interacted.

Central Eurasians positioned themselves between or among existing elements of different civilisational contacts, both nomadic and sedentary. Yet they were able to retain properties and perceptions of themselves and the other in a stable yet flexible manner to create robust societies. This meant that society could function with a ruling elite of Persian and later Turkic descent over diverse ethnic populations, use Persian in government and Arabic in formal religious settings, and showcase power and leadership in specific ways. Consequently, Central Eurasian societies exhibited layered understandings and practices in line with strongly interacting elements. This shared ecosystem, where each element and society impacted governance proved more efficient, and thus reliable. In other words, different groups in wider Eurasia could continue to function in familiar ways whether sedentary or nomadic, and yet have spaces of interaction that improved rather than impeded development. For example, Central Eurasians were integral to the functioning of the thirteenth century Mongol Empire in Eurasia as traders, governors and bureaucrats. They were able to speak multiple languages and exist in a wide variety of spaces and synthesised what we call today a complex adaptive system exhibiting strategies that proved resilient for them and for their communities. They created capacity in order to fashion positions of influence in government, business and in the maintenance of a vibrant cultural milieu that came to simultaneously represent pre-Islamic, Islamic, Turkic as well as Mongol in medieval Eurasia (Cosmo et al. 2009). As a result, Central Eurasia functioned as the host society within which strongly interacting elements, such as different religions, ethnic groups and races, found resonance and contributed to instilling meaning and significance.

This was somewhat disrupted by Tsarist Russia when it introduced a systematised and bureaucratised government in Central Eurasia beginning in the nineteenth century. The Directorate of Turkestan began to transform a society based on Islamic law, deeply integrated with the cultural principles of society and the place of the individual within it. The move away from personalised relationships and cultural contexts especially in small Muslim communities like those in the Indian subcontinent and Central Eurasia created ruptures in state-citizen relationships precisely because the local milieu was dismissed as traditional, and thus redundant. However, Central Eurasian societies found avenues of belongings, for example through participation in Sufi orders. Often Sufi saints performed the role of facilitators and translators of this new modernity. With their help, Muslim societies were able to adapt to the changing modes of communication, education, knowledge production and lived experience. These Sufi orders were able to bridge the gap between a society based on personalistic relationships and an impersonal automated European state, whose main aim was control (Leibeskind 1998, 12, 28). This same relationship was reproduced in the Soviet Union where, for example, citizens of Uzbek SSR learned to be Uzbek and Soviet, simultaneously finding a balance by establishing new meanings of belonging with the help of writers and intellectuals (Hirsch 2000; Shin 2015; Khalid 2015). These interactions imbued their own logic, language and modes of understanding which are still visible in modern Central Eurasian states today confirming how strongly interacting elements continue to be performative in these societies. While Chandler (2014, 49) refers to this as producing order through self-organisation on the basis of interactive processes, Wendt (1992, 407) refers to these as collective meanings that are always in process but are constitutive of identity and interests.

Eurasian reconnect and emergent coherence

Central Eurasian society, well versed with polyglot and multi-ethnic empires of the past from the Xiong-nu, Mongols and subsequent Khanates, holds within its historical experience an arsenal of repositories of practice that 'act as locus around which varied performances of power politics are created' (Neumann and Wigen 2018, 9). More specifically, the role played by Central

Eurasian societies and individuals in the administration and economy in the medieval period made them the real movers and shakers of the Eurasian world and is illustrative of their resilience. They acted as ministers, governors, administrators, traders and translators for the ruling nomadic elites and provided the interface with neighbouring sedentary polities. As traders, they had access to dense networks spread across the Eurasian region important not only for economic reasons but also for reasons of conquest and expansion (Kalra 2020b). Having knowledge of diverse societies coupled with information about trade routes and relationships stretching back centuries, Central Eurasian communities served as the nodes of contact that nomadic empires *and* sedentary polities relied on for exchange. This not only guaranteed access to places of honour and influence in court but represented leverage and real agency. Partnerships between rulers and their Central Eurasian consultants and administrators reached an apex in the Mongol Empire of exchange (Kotkin 2007). What followed were intensely strong interactions (intercalation) that were natural breeding grounds for the emergence of a complex adaptive system and resilience-thinking that could withstand shock and still flourish.

The main element to be emphasised here is the importance given to connectivity that ensured mobility, exchange and access to a wide array of resources. This brings us to the strategy that we refer to as the desire for connectedness that is the building block for resilient communities in this region, one that has been founded on the principles of the processes of intercalation. The connectedness of Central Eurasian communities, physical, economic and social, implies a shared space that underscores meaning and understanding. Scott Levi weaves together a narrative of the shared history of Central Eurasia as connectivity that persisted even in the face of European colonisation and the perceived decline of the Silk Road(s) (Levi 2020). The economies of Central Eurasia and trade in the region continued to flourish even when maritime trade was seen as usurping the bulk of trade in the world. Levi cites the exchange of information especially in matters of trade and commerce along with the continued support of physical and institutional infrastructure (financing of projects and provision of security) and other policies that could be seen across the wider Eurasian space: Central Eurasia, Iran and even India in the sixteenth-seventeenth centuries (Levi 2020, 77–78). The tools of understanding Central Eurasia need to accommodate these medieval empires (nomadic) and their partners (Central Eurasian polities) to understand the importance of connectivity in the region.

In addition, the lived experience of the Soviet Union also determined behaviour and practice towards keeping the region connected. The Soviet Union standardised national infrastructure and set up transnational infrastructure, economic relations and ways of behaving that have outlasted dissolution, albeit with disruptions (Kotkin 2007, 525). While the rules of engagement underwent a transformation, the borders of Central Eurasian nation-states and the nationalisms encouraged in Soviet times continue to inform communities and state actors in the region. The identity and behaviours associated with the republics played out later as these countries became sovereign nations with minor fluctuations (Kudaibergenova 2020). There was also an almost immediate attempt at reconnecting the region after 1991, which showcases cooperation as a strategy for resilience in the region. Specifically, in the modern era and the current world order, the nation-state is the main actor that exercises the will of the people and thus its participation in macro level regional organisations is of interest here.

Regional cooperation in the modern era

How do we understand the emergence of a number of cooperative organisations in the region? Theories of international relations applied to post-Soviet Central Eurasia have too often used rationalism and/or reflectivism to explain behaviour and practice (Kalra and Saxena 2007; Dadabaev and Heathershaw 2020). One of the more comprehensive explanations has been the category of 'protective integration'. Allison applies the term to explain the motivations of Central Eurasian states joining organisations like SCO and CSTO. These motivations include primarily the means to enhance sovereignty, regime security, stability and legitimacy. The continued participation in meetings of collective organisations in the region provides what social constructivists call normative-framework building opportunities (Allison 2018, 298). Participation in these organisations justify and legitimate domestic political practices and sustain a conservative sovereignty-focused normative framework. They build and reinforce collective identities through socialisation practices in these meetings and in doing so confirm bonds of 'statist, sovereignty focused norms' (Allison 2018, 309). However, this application of protective integration from a security perspective, while acknowledging the support of such macro level organisations for norm making behaviour and practice for security purposes, fails to see how these organisations serve as avenues for negotiation that take into account local needs (economic and social) at the same time. The increasingly complex world poses severe challenges to livelihood, especially when means of exchanges that are so important for these communities are truncated due to international borders, undue customs regimes or state level threats. In these circumstances, it is essential to highlight interactions in the past where these communities have tended toward cooperation and collaboration. In doing so, we reiterate the ways in which this region has withstood changes and come out resilient. It locates the basis of resilient thinking in historic interactions that unfolded on the large expanse of Eurasia over millennia.

Eurasian economic union and the BRI

At this juncture we must turn to the most recent of such initiatives: the Eurasian Economic Union (EAEU) and the Belt and Road Initiative (BRI). Much has been said about the two initiatives in scholarly literature, from positioning them as anti-Western, hegemonic, establishing spheres of influence and linking up with Europe and Asia-Pacific in the longer term and in the international global arena. The EAEU has been seen as Moscow's attempt at creating a Greater Europe (Vinokurov 2014); bridging economic and political ideals (Popescu 2014); regional integration leading to external borders (Safranchuk 2015); President Vladimir Putin's project, 'pole' in international politics, a historic moment challenging the unipolar world order with civilisational contacts as the bedrock of coming together. Similarly, the BRI has been called a tool of soft power and diplomacy (Kratz 2015; Shambaugh 2015); China's external economic policy (Callahan 2016; Ferdinand 2016; Summers 2016); a means for win-win cooperation (Wang 2015; Liu 2015); President Xi Jinping's assertion in China's domestic policy; a reassertion of the traditional Silk Road and China's Marshall Plan. In terms of the influence that Russia and China have on Central Eurasia, the EAEU and the BRI are the predominant frames of analysis at present. Scholars have opined that the EAEU and BRI are being used to create spheres of influence, one with a spatial (Russia) logic, and the other with a functional (China) logic (Kaczmarski 2017, 1040).

The analysis, however, remains welded to the big powers and the lack of agency on the part of Central Eurasia is palpable. Kazakhstan and even Azerbaijan are reduced to being essentially connecting points between Europe and China (Kaczmarski 2017, 1038) bringing to bear the mistakes of the past (Levi 2020; Adibayeva and Dadabaeva 2013). While it is easy to see that Russia and China are positioned to benefit from these developments, it is essential to acknowledge that Central Eurasian countries have consistently entered on a voluntarily basis and more significantly initiated regional organisations and groupings which, in their own words, give them an equal platform, enable access to international markets, and are places to renegotiate their relationships with their neighbours (Vanderhill et al. 2020; Dadabaev and Heathershaw 2020). However, the current analysis and scholarly engagement with the region continues to overwhelmingly represent Central Eurasian countries as trapped and having little to no agency. Whether it is the New Great Game playing out between the US and China, or the resurgence of the concept of the Silk Road(s), Central Eurasia remains on the periphery even when it is the object of interest.

Seen from a historic perspective, the BRI offers bilateral and multilateral interaction on the historic Silk Road(s) and reflects connectivity in the region which has stretched back centuries. It is premised on the notion of comprehensive development for Eurasia, which has the potential to benefit China along with the other countries along the Silk Road(s) (Kalra 2019). This needs to be viewed in conjunction with how cooperation based on inclusivity is the cornerstone of foreign policy directives of countries in the region. For example, Kazakhstan's multi-vector foreign policy and Turkmenistan's policy of neutrality (Mammedov 2020) are both examples of Central Eurasian countries finding their own language of inclusiveness to engage with the outside world. They are both premised on inclusivity, not exclusivity by their own admission. However, the civilisational conflict between Russia and the so-called West (Europe, UK and the US) and/or now China and the West make Central Eurasian countries the real losers. No matter who actually proposes an idea, it is seen as coming from without even in the face of facts and empirical evidence. In 1994, the First President of Kazakhstan, N. Nazarbayev, suggested the creation of a Eurasian Union with freedom of movement and free trade across the Eurasian space. Nazarbayev during his time as President (1990–2019) continued to express Eurasian ideas with Astana (now Nursultan), not Moscow as the centre (Kalra & Saxena 2015). However, the Eurasian Economic Union is seen *only* as a Russian (read President Vladimir Putin's) project (Kalra 2020b). It is important to note that President Putin's (2000–2008; 2012–Current) pivot towards Asia and/or 'Look East' policy was announced as late as 2010 (Bakare 2021; Sakwa 2018). Similarly, in the case of the BRI,

President Xi Jinping's announcement of the BRI in Kazakhstan in 2015 is mentioned in passing and is widely seen as another attempt at garnering Chinese influence with little to no regard for Kazakhstan's agency in this process.

Central Eurasian agency

The inability to see the countries of Central Eurasia as countries with their own interests and motivations as driving forces makes Kazakhstan, Uzbekistan and all the other countries of Central Eurasia objects to be acted upon. The reality on the ground is that the economies of Central Eurasia are institutionally and infrastructurally connected with Russia and it serves as an important labour and consumer market destination, while China is the biggest economic partner in the region since the early 2000s and a major investor in infrastructure projects. With the oil pipeline connecting China-Kazakhstan-Uzbekistan-Turkmenistan, exit routes from the region have diversified to go beyond relying solely on Russia, and there is further investment in roads, land ports and other land infrastructure to expand trade in Eurasia from which the Central Eurasian countries will also benefit (Kazczmarski 2017, 1036). The interconnections between and with the economies in the region are as important to domestic economies for their own sake, as for the larger global network (Kuchins et al. 2015). As in medieval times, one of the main reasons for the unprecedented success of the Mongols (thirteenth century) was infrastructure support, institutional and physical, that linked micro economies (carpet making, textiles, pottery) of Central Eurasia to the rest of the known world at the time (Abu-Lughod 1989; Marsden & Hopkins 2019; Kalra 2018; 2020b). Levi (2020), Di Cosmo (1994), CiociLtan (2012), Prazniak (2010), Kolbas (2006), Martinez (2009) and others speak of the connectedness of the Eurasian space as a forerunner to prosperity and influence. They report that 'the Mongols integrated Eurasia on an unprecedented scale, a globalisation of the Old World that contributed to the discovery of the New World and helped shape the early modern period' (Biran 2015, 535). In this endeavour, the role of Central Eurasian traders and administrators cannot be overstated. They had the knowledge of polities, societies and trade networks necessary for this economic transformation. The Mongol trade juggernaut was no less an outcome of Central Eurasian ingenuity than the steppe tradition brought by the Mongol Khans (Kalra 2020a). The unique geographical location inhabited by Central Eurasians allowed them to function as interlocutors, mediators and translators and connected the vastness of the Eurasian space. Relationships thus formed between Central Eurasia and their neighbouring Russian and Chinese communities date back centuries and continue to be important and necessary.

Wendt's approach to identity formation and his use of the concept of intentional transformation applies to the countries of Central Eurasia precisely because it can identify the nation-states in the region through the lens of their own (not European) historical path dependence. It locates the elements of community and state locally, while at the same time acknowledging a cooperative world order as an end goal. These two assumptions are in line with the historical understanding of the region whereby Central Eurasia has withstood a number of shocks and found that connectivity introduces flows into a system that help it rebalance under threat. In the context of the VUCA world and consequent resilience thinking, it becomes imperative to refer to the natural proclivity of small communities (Central Eurasia) inhabiting a vast expanse (Eurasia) to find the basis of identity, behaviour and practice, well within the paradigm of cooperation.

Inherent resilience and the process of intercalation

The Eurasian narrative with medieval empires at the centre is a story unlike one told from a Sinocentric, Russocentric or Eurocentric perspective. At its heart are the intersections of Turkic, Iranian and other peoples of Central Eurasia who have inhabited diverse ecological environments and see connectivity as a benefit not an impediment or threat. These strategies have been centuries in the making and have at their core inter-civilisational contact in all spheres: political, economic and social. Beginning with the nomadic, semi-sedentary, and sedentary civilisations that included contact with Chinese, Persian, and Islamic worlds, Turko-Mongol empires, European colonial powers, the Tsarist Russian Empire and ending with the Soviet Union. The countries of Central Eurasia have lived through multiple world systems and withstood shocks, unpredictability and challenges which can be referred to as multiple VUCA worlds. Central Eurasia was the place where steppe traditions met with sedentary traditions (China, pre-Islamic Persia) to be translated, acculturated, modified and even codified (McChesney 2009, 278; Rouse 2020, 399–400). These spaces and communities served as the nodal points linking disparate civilisations and became zones for transmission of cultural values that came to represent their in-between nature and the accommodation of a diverse set of practices. Rather than disappearing into the east or the west, Europe or Asia, or any such seemingly opposing categories, Central Eurasia has within its history the tools to operate in all of them (Gleason 2003, 3). These communities have developed erudite ways through their exposure to strongly interacting systems (intercalation) that have created layered understandings of themselves and the other. They have performed and practiced resilience through these interactions that accommodate the past to curate a connected future. This requires immediate recognition because it came about through centuries of interactions, negotiations and exchanges. Even in the modern era, as part of the Soviet Union, Central Eurasia represented a place where different ideologies (political and economic) collided, yet found compromise. In terms of resilient communities and the larger global conversations concerning international orders and debates about decolonisation, Central Eurasian countries serve as noteworthy examples. They can either be understood from the perspective that finally recognises their role(s) in world history or they can continue to be ignored to perpetuate great power politics and ensuing misunderstandings.

The building blocks of the Eurasian narrative are found in the expanse of empires, not the confines of nation-states. Central Eurasian states and communities continue to be resilient mainly through reconnecting and finding spaces for interactions that overcome differences. Connectivity has served Central Eurasia well in their past encounters with multiple *others* (Chinese, Persians, Arabs, Mongols and even Russians). If in the medieval period, traders and merchants proved to be successful translators in fashioning a Turko-Mongol empire of great purport (Kalra 2020a), in the colonial period, there were instances of Sufi saints and leaders facilitating and translating European modernity that challenged the traditional relationship between state and citizen (Leibeskind 1998, 12, 28). These examples point to the capacity within Central Eurasia to find those interlocutors time and again to help navigate external shocks and come out resilient and stronger. In the modern era, these interlocutors exist at various levels in the community and at state level. They find those *connectivities* whether through participating in macro level organisations, building physical and institutional infrastructure, or community interactions and exchanges that in many cases represent the entirety of economic functions locally. These represent the ways in which intercalation processes unfolded in Central Eurasia and how these societies have time and again played host to strongly interacting elements.

To return to the idea of resilience as a quality and an analytic of governance in the case of Central Eurasia means to use the category of connectedness beyond territoriality and sovereignty (Wendt 1992, 402) in understanding the region. A connectedness that emerged out of geography and history in the first place is visible today in the modern era through participation in collective initiatives and organisations. It is one of the more reliable means to instil meaning in behaviour and practice of Eurasian polities in view of the historical trajectory of the region. Unlike the homogenous state system that Waltz (1979) believed was the international arena of the modern era, it is the diversity within the global order that needs acknowledgement today, with a view towards accepting a variety of strong states operating together in a shared space. Concepts of good life and resilience are, at their core, contested because the 'local' can never be homogenous. However, what is increasingly clear is that the understanding of the various 'locals' that together make-up the 'global' is the need of the hour. With this in mind, this article hopes to leave the reader with a better understanding of the necessity of different categories of analysis in the pursuit of understanding how Eurasian communities underwent processes of intercalation and carry within them an inherent resilience.

Acknowledgements

This research has been made possible thanks to the GCRF COMPASS project (ES/P010849/1). This paper would not be possible without the insightful comments and constant support (intellectual and otherwise) of Dr Siddharth S Saxena. The editors of this special issue and reviewers deserve a special note of gratitude for their feedback which was invaluable in conveying the key ideas here.

Disclosure statement

No potential conflict of interest was reported by the author(s).

ORCID

Prajakti Kalra ⓘ http://orcid.org/0000-0002-6193-7468

References

Abu-Lughod, J. 1989. *Before European Hegemony: The World System A.D.1250–1350/Janet L. Abu-Lughod*. New York, NY: Oxford University Press.

Adibayeva, A., and G. Dadabayeva. 2013. Mackinder's Legacy Today. In *Central Asia in International Relations*, edited by N. Megoran and S. Sharapova, *Central Asia in International Relations*. Oxford: Oxford University Press.

Akçali, E. and M. Perinçek. 2009. "Kemalist Eurasianism: An Emerging Geopolitical Discourse in Turkey." *Geopolitics* 14 (3): 550–569.

Allison, R. 2018. "Protective Integration and Security Policy Coordination: Comparing the SCO and CSTO." *The Chinese Journal of International Politics* 11 (3): 297–338.

Bakare, N. (2020). "Contextualizing Russia and South Asia Relations through Putin's Look East Policy." *Journal of Asian and African Studies (Leiden)* 56 (3): 676–92.

Bennis, W., and B. Nanus. 1986. *Leaders: The Strategies for Taking Charge (Perennial library)*. New York, NY: Harper & Row.

Biran, M. 2015. "The Mongol Empire and Inter-Civilizational Exchange." In *The Cambridge World History*, Vol. 5, The Cambridge World History, 534–558. Cambridge: Cambridge U-niversity Press.

Bousquet, A. and S. Curtis. 2011. "Beyond Models and Metaphors: Complexity Theory, Systems Thinking and International Relations." *Cambridge Review of International Affairs* 24 (01): 43–62.

Callahan, W. 2016. "China's "Asia Dream" The Belt Road Initiative and the New Regional Order". *Asian Journal of Comparative Politics* 1 (3): 226–43.

Canfield, R. L., and School of American Research. 1991. *Turko-Persia in Historical Perspective*. School of American Research Advanced Seminar Series. Cambridge: Cambridge University Press.

Chandler, D. 2010. *International Statebuilding: The Rise of Post-liberal Governance*, Vol. 2. Critical Issues in Global Politics. London: Routledge.

Chandler, D. 2014. "Beyond Neoliberalism: Resilience, the New Art of Governing Complexity." *Resilience* 2 (1): 47–63.

Christian, D. 1994. "Inner Eurasia as a Unit of World History." *Journal of World History* 5 (2): 173–211.

Ciocîltan, V. 2012. *The Mongols and the Black Sea Trade in the Thirteenth and Fourteenth Centuries (East Central and Eastern Europe in the Middle Ages, 450–1450, 1872–8103; v. 20)*. Boston: Brill.

Cowan, P. J. 2007. "Geographic Usage of the Terms Middle Asia and Central Asia." *Journal of Arid Environments* 69 (2): 359–63.

Dadabaev, T., and Heathershaw, J. 2020. "Central Asia: A Decolonial Perspective on Peaceful Change" In *The Oxford Handbook of Peaceful Change in International Relations*, edited by T. V. Paul, Deborah Welch Larson, Harold A. Trinkunas, Anders Wivel, and Ralf Emmers. Oxford: Oxford University Press.

Di Cosmo, N. 1994. "Ancient Inner Asian Nomads: Their Economic Basis and Its Significance in Chinese History." *The Journal of Asian Studies* 53 (4): 1092–126.

Di Cosmo, N. 1999. "The Northern Frontier in Pre–Imperial China." In *The Cambridge History of Ancient China*. The Cambridge History of Ancient China, 885–966. Cambridge: Cambridge University Press.

Di Cosmo, N., A. Frank, and P. Golden. 2009. *The Cambridge History of Inner Asia: The Chinggisid Age*. Cambridge: Cambridge University Press.

Ferdinand, P. 2016. "Westward ho—The China Dream and "One Belt, One Road": Chinese Foreign Policy under Xi Jinping" *International Affairs* 92 (4): 941–57.

Frank, A. G. 1992. "The Centrality of Central Asia." *Studies in History (Sahibabad)* 8 (1): 43–97.

Fukuyama, F. 1992. *The End of History and the Last Man*. London: Hamish Hamilton.

Gleason, G. 2003. "Centrality of Central Eurasia." *Central Eurasian Studies Review* 2 (1): 2–6.

Goody, J. 2006. *Theft of History*. New York, NY: Cambridge University Press.

Hirsch, F. 2000. "Toward an Empire of Nations: Border-Making and the Formation of Soviet National Identities." *The Russian Review (Stanford)* 59 (2): 201–26.

Hirsch, F. 2005. *Empire of Nations: Ethnographic Knowledge and the Making of the Soviet Union*. Ithaca, NY: Cornell University Press.

Holling, C. S. 2001. "Understanding the Complexity of Economic, Ecological, and Social Systems." *Ecosystems (New York)* 4 (5): 390–405.

Huntington, S. P. 2011. *The Clash of Civilizations and the Remaking of World Order*. New York, NY: Simon & Schuster.

Ikenberry, G. J. 2005. "Power and Liberal Order: America's Postwar World Order in Transition." *International Relations of the Asia-Pacific* 5 (2): 133–52.

Jackson, J. 2012–2018. Earthquakes Without Frontiers Project, a joint NERC-ESRC consortium. http://ewf.nerc.ac.uk/ewf-projects/central-asia/

Kaczmarski, M. 2017. "Non-western Visions of Regionalism: China's New Silk Road and Russia's Eurasian Economic Union." *International Affairs (London)* 93 (6): 1357–76.

Kalra, P. 2018. *The Silk Road and the Political Economy of the Mongol Empire*. Routledge studies on the Chinese economy; 68. London: Routledge.

Kalra, P. 2019. "Development Paradigms in Eurasia: The Case of BRI." Paper presented at the 'The Belt and Road Initiative @Five' and 'BRI and the Silk Road' Workshop, March 13–15. Kalra, P. and E. Van Gils. eds. 2019. https://www.kent.ac.uk/politics/rs-gcrf-compass/COMPASS-BRI-PROCEEDINGS.june19-for-printing.pdf. 20–23.

Kalra, P. 2020a. "Pax Mongolica: Trade and Traders in the Mongol Empire." In *Oxford Research Encyclopedia of Asian History*. Oxford: Oxford University Press.

Kalra, P. 2020b. "Is Central Asian History Peripheral?" https://www.peripheralhistories.co.uk/post/is-central-asian-history-peripheral.

Kalra, P. and S. S. Saxena. 2007. "Shanghai Cooperation Organisation and Prospects of Development in the Eurasia Region." *Turkish Policy Quarterly (TPQ)* 6 (2): 95–99.

Kalra, P., and S. S. Saxena. 2015. "The Asiatic Roots and Rootedness of the Eurasian Project." In *The Eurasian Project and Europe*, edited by D. Lane and V. Samokhvalov, 38–52. London: Palgrave Macmillan.

Kalra, P., and S. S. Saxena. 2021. "Globalising Local Understanding of Fragility in Eurasia." Saxena, S. S. and P. Kalra eds (2021). *Journal of Eurasian Studies*. Hanyang Taehakkyo. A-T'ae Chiyŏk Yŏn'gu Sent'ŏ, issuing body. Produced and distributed by SAGE for the Asia-Pacific Research Center of Hanyang University.

Kavalski, E. 2007. "Partnership or Rivalry between the EU, China and India in Central Asia: The Normative Power of Regional Actors with Global Aspirations." *European Law Journal: Review of European Law in Context* 13 (6): 839–856.

Keohane, R. 2005. *After Hegemony Cooperation and Discord in the World Political Economy*, 1st Princeton classic ed., Princeton classic editions. Princeton, NJ: Princeton University Press.

Khalid, A. 2000. "Russia, Central Asia and the Caucasus to 1917." In *The New Cambridge History of Islam*, Vol. 5, 180–202. Cambridge: Cambridge University Press.

Khalid, A. 2015. Making Uzbekistan: Nation, Empire, and Revolution in the Early USSR.

Khazanov, A., and J. Crookenden. 1986. *Nomads and the Outside World*. Cambridge studies in social anthropology; 44. Cambridge: Cambridge University Press.

Kolbas, J. 2006. *The Mongols in Iran: Chingiz Khan to Uljaytu 1220–1309*. London: Routledge.

Korosteleva, E. and I. Petrova. 2022. "What Makes Communities Resilient in Times of Complexity and Change?" *Cambridge Review of International Affairs* 35 (2): 137–157.

Korosteleva, E., and T. Flockhart. eds. 2020. *Resilience in EU and International Institutions: Redefining Local Ownership in a New Global Governance Agenda*, 1st. ed. Routledge.

Kotkin, S. 2007. "Mongol Commonwealth?: Exchange and Governance across the Post-Mongol Space." *Kritika (Bloomington, Ind.)* 8 (3): 487–531.

Kratz, A. 2015. "China's AIIB: A Triumph in Public Diplomacy." In *One Belt, One Road': China's Great Leap Outward*, edited by F. Godement and A. Kratz. Paris: European Council on Foreign Relations.

Kuchins, A., J. Mankoff, A. Kourmanova, O. Backes. 2015. *Central Asia in a Reconnecting Eurasia. Uzbekistan's Evolving Foreign Economic and Security Interests (CSIS)*. New York, NY: Rowman and Littlefield.

Kudaibergenova, D. T. 2020. *Toward Nationalizing Regimes. Conceptualizing Power and Identity in the Post-Soviet Realm*. Pittsburgh, PA.: University of Pittsburgh Press.

Liebeskind, C. 1998. *Piety on Its Knees: Three Sufi Traditions in South Asia in Modern Times*. Oxford: Oxford University Press.

Levi, S. 2020. *The Bukharan Crisis: A Connected History of 18th-Century Central Asia (Central Eurasia in context)*. Pittsburgh, PA: University of Pittsburgh Press.

Liu, Y. 2015. ""Marshall Plan" Copycat Allegations Misleading." *Beijing Review*, No. 6, 5 February.

Mammedov, N. 2020. Cultural Foundations of Turkmenistan's Permanent Neutrality. Address in Panel Discussion on 25 Years of Turkmenistan's Neutrality in Cambridge on 6 November 2020.

Marsden, M., and B. Hopkins. 2019, May 23. "Afghan Trading Networks." Oxford Research Encyclopedia of Asian History. Retrieved November 12, 2020, from https://oxfordre.com/asianhistory/view/10.1093/acrefore/9780190277727.001.0001/acrefore-9780190277727-e-119.

Martin, T. 2011. *The Affirmative Action Empire*. Ithaca, NY: Cornell University Press.

Martinez, A. P. 2009. "Institutional Development, Revenues and Trade." In *The Cambridge History of Inner Asia*, pp. 89–108. Cambridge: Cambridge University Press.

McChesney, R. 2009. "The Chinggisid Restoration in Central Asia: 1500–1785." In *The Cambridge History of Inner Asia*, edited by N. Di Cosmo, A. J. Frank and P. B. Golden, 277–302. Cambridge: Cambridge University Press.

Mearsheimer, J. 2018. *The Great Delusion: Liberal Dreams and International Realities*. Henry L. Stimson lectures. London: Yale University.

Megoran, N., and S. Sharapova. 2013. *Central Asia in International Relations*. Oxford: Oxford University Press.

Milner, H. 1992. "International Theories of Cooperation: Strengths and Weaknesses." *World Politics* 44: 466–496

Mostafa, G. 2013. "The Concept of 'Eurasia': Kazakhstan's Eurasian Policy and its Implications." *Journal of Eurasian Studies* 4 (2): 160–170.

Murashkin, N. 2018. "Not-so-New Silk Roads: Japan's Foreign Policies on Asian Connectivity Infrastructure Under the Radar." *Australian Journal of International Affairs* 72 (5): 455–472.

Neumann, I., and E. Wigen. 2018. *The Steppe Tradition in International Relations: Russians, Turks and European State Building 4000 BCE–2018 CE*. Cambridge: Cambridge University Press.

Popescu, N. 2014. *Eurasian Union: The Real, the Imaginary and the Likely*, Working Paper 132. Paris: Institute for Security Studies.

Prazniak, R. 2010. "Siena on the Silk Roads: Ambrogio Lorenzetti and the Mongol Global Century, 1250–1350." *Journal of World History* 21 (2): 177–217.

Robinson, F. 1997. "Ottomans-Safavids-Mughals: Shared Knowledge and Connective Systems." *Journal of Islamic Studies (Oxford, England)* 8 (2): 151–184.

Robinson, F. 2007. "The Mughal Dynasties: Francis Robinson Looks for the Distinctively Tolerant and Worldly Features of Mughal Rule in India and that of the Related Islamic Dynasties of Iran and Central Asia." The Free Library. History Today Ltd. 18 Dec. 2020. https://www.thefreelibrary.com/The+Mughal+dynasties%3a+Francis+Robinson+looks+for+the+distinctively...-a0165362660

Robinson, F. ed. 2010. *The New Cambridge History of Islam*, Vol. 5. The New Cambridge History of Islam. Cambridge: Cambridge University Press.

Rouse, L. M. 2020. "Silent Partners: Archaeological Insights on Mobility, Interaction and Civilization in Central Asia's Past." *Central Asian Survey* 39 (3): 398–419.

Roy, O. 2000. *The New Central Asia. The Creation of Nations. Library of international relations (Series); 15.* London: Tauris.

Safranchuk, I. 2015. "Globalizatsiya v golovakh." *Rossiya v global'noi politike* 14: 1.

Sakwa, R. ed. 2018. *Eurasia on the Edge: Managing Complexity,* edited by Richard Sakwa.

Saxena, S. S. 2019. 'Preface." In *Making the Kyrgyz,* edited by P. Kalra and J. Waddle. Cambridge: Cambridge Scientific Publishers.

Shambaugh, D. 2015. "China's Soft-Power Push: The Search for Respect." *Foreign Affairs (New York, N.Y.)* 94 (4): 99–107.

Shin, B. 2015. *Between the Uzbek and the Soviet: Uzbek Identity Construction through Soviet Culture from the 1930s to 1940s.* University of Cambridge. Department of Slavonic Studies, degree granting institution.

Sneath, D. 2007. *The Headless State: Aristocratic Orders, Kinship Society, and Misrepresentations of Nomadic Inner Asia.* New York, NY: Columbia University Press.

Summers, T. 2016. 'China's "New Silk Roads": Sub-National Regions and Networks of Global Political Economy.' *Third World Quarterly* 37 (9): 1628–1643

Vanderhill, R, S. F. Joireman, and R. Tulepbayeva. 2020. "Between the Bear and the Dragon: Multivectorism in Kazakhstan as a Model Strategy for Secondary Powers." *International Affairs (London)* 96 (4): 975–993.

Vieira, M. A. 2016. "Understanding Resilience in International Relations: The Non-Aligned Movement and Ontological Security." *International Studies Review* 18 (2): 290–311

Vinokurov, E. 2014. "From Lisbon to Hanoi: The European Union and the Eurasian Economic Union in Greater Eurasia." In *Russia's 'pivot' to Eurasia,* edited by Liik, K. London, European Council on Foreign Relations.

Walker, B., D. Salt, and W. Reid. 2006. *Resilience Thinking Sustaining Ecosystems and People in a Changing World.* Washington, DC: Island Press.

Wang, S. 2015. "Commentary: Chinese Marshall Plan Analogy Reveals Ignorance, Ulterior Intentions." Xinhua, 11 March.

Waltz, K. (1979). *Theory of international politics/Kenneth N. Waltz (University of California, Berkeley).* Addison-Wesley Series in Political Science. New York, NY: McGraw-Hill.

Weller, Thomas E., Mark Ellerby, Siddharth S Saxena, Robert P Smith, and Neal T Skipper. 2005. "Superconductivity in the Intercalated Graphite Compounds C6Yb and C6Ca." *Nature Physics* 1 (1): 39–41.

Wendt, A. 1992. "Anarchy is What States Makes of It: The Social Construction of Power Politics." *International Organization* 46 (2): 391–425.

Index

Italicized and **bold** pages refer to figures and tables respectively, and page numbers followed by "n" refer to notes.

For Product Safety Concerns and Information please contact our EU
representative GPSR@taylorandfrancis.com
Taylor & Francis Verlag GmbH, Kaufingerstraße 24, 80331 München, Germany

www.ingramcontent.com/pod-product-compliance
Lightning Source LLC
Chambersburg PA
CBHW060559060326
40690CB00017B/3762